THE “MENTAL” AND THE “PHYSICAL”

THE ESSAY "The 'Mental' and the 'Physical'" first appeared in Volume II of *Minnesota Studies in the Philosophy of Science: Concepts, Theories, and the Mind-Body Problem*, edited by Herbert Feigl, Michael Scriven, and Grover Maxwell and published by the University of Minnesota Press in 1958.

THE "MENTAL" AND THE "PHYSICAL"

The Essay and a Postscript

HERBERT FEIGL

University of Minnesota Press, Minneapolis

Printed in the United States of America

Library of Congress Catalog Card Number: 67-24556

PUBLISHED IN GREAT BRITAIN, INDIA, AND PAKISTAN BY THE OXFORD UNIVERSITY PRESS, LONDON, BOMBAY, AND KARACHI, AND IN CANADA BY THE COPP CLARK PUBLISHING CO. LIMITED, TORONTO

Preface to Essay

I feel I should acknowledge my sincere indebtedness to the countless philosophers and scientists who have helped me by their publications as well as (in many instances) by personal discussion or correspondence to reach whatever clarity I may claim to have achieved. It is impossible to mention them all, but some stand out so distinctly and prominently that I should list them. Naturally, I have learned from many of these thinkers by way of disagreement and controversy. In any case none of them is to be held responsible for whatever may be wrong or confused in my views. My first acquaintance with philosophical monism goes back to reading the work of Alois Riehl; I found essentially the same position again in Moritz Schlick, some of whose work I had studied before I became his student in Vienna in 1922. I have profited enormously (although *he* may well think, not sufficiently) from discussions with my kind and patient friend R. Carnap intermittently throughout more than thirty years. During my Vienna years (1922–30) I was greatly stimulated by discussions also with Schlick, Wittgenstein, Victor Kraft, Otto Neurath, E. Kaila, Karl Popper, Edgar Zilsel, *et al.* I was greatly reinforced in my views by my early contact with the outstanding American critical realist C. A. Strong (in Fiesole, Italy, 1927 and 1928). Along similar lines I found corroboration in the work of Roy W. Sellars, Durant Drake, and Richard Gätschenberger, and in some of the writings of Bertrand Russell. Discussions (and many controversies) during my American years, beginning in 1930, with E. G. Boring, S. S. Stevens, P. W. Bridgman, C. I. Lewis, A. N. Whitehead, H. M. Sheffer, V. C. Aldrich, S. C. Pepper, E. C. Tolman, C. L. Hull, B. F. Skinner, K. Lewin, E. Brunswik, W. Köhler, Albert Einstein, H. Reichenbach, F. C. S. Northrop, and Philipp Frank proved most stimulating.

During the last three and a half years of the activities of the Minnesota Center for Philosophy of Science not only did I have the tremendous advantage of intensive discussions with my colleagues Paul E. Meehl, Wilfrid Sellars, and Michael Scriven, each of whom disagrees with me on several different fundamental points, and each for different reasons, but I also profited from discussions with such visitors or collaborators as Gilbert Ryle, C. D. Broad, Anthony Flew, Peter Strawson, Ernest Nagel, C. G. Hempel, A. Kaplan, Arthur Pap, Herbert Bohnert, Henry Mehlberg, Hilary Putnam, Gavin Alexander, William Rozeboom, and Adolf Grünbaum. Last, but not least, I owe a great debt of gratitude to my students at Minnesota who during many a year of seminar work in the philosophical problems of psychology have helped me through their criticisms to arrive at clearer formulations of my ideas and to eliminate various difficulties, mistakes, and confusions. It has been a veritable Odyssey of ideas for me, and I am by no means sure I have "arrived"!

HERBERT FEIGL, Director,
Minnesota Center for Philosophy of Science

February 1957

Preface to Postscript

The surprisingly great number of requests for reprints of my essay of 1958, and the friendly insistence of John Ervin, director of the University of Minnesota Press, finally made me decide in favor of republishing it as a separate monograph. The appended "Postscript after Ten Years" will, I hope, make clear that I have had second and third thoughts—as well as "grayer mornings"—on the Identity Theory. As I indicate in the "Postscript," I still consider the basic intent of a monistic view scientifically plausible, and philosophically (logically) acceptable—provided it is reformulated in a more careful way. My own reformulation is thus far only a "blueprint." May I again invite proponents and opponents to help with constructive as well as destructive criticism?

My indebtedness to other thinkers is very great again. If I were to single out those who have helped me most directly and intensively, I would mention first my friends Rudolf Carnap, Paul Feyerabend, Grover Maxwell, Paul E. Meehl, and Wilfrid Sellars. But there are many others—some like J. J. C. Smart, D. M. Armstrong, Brian Medlin, and John Passmore in Australia; several of my graduate students (notably, William Demopoulos and Henry Lackner); and most recently and prominently, Professor Keith Gunderson and Dr. Judith Economos (both at UCLA) who have helped me enormously by incisive and pertinent criticism. I am equally grateful to Professors Bruce Aune and John Kekes who, from differing points of view, have discussed with me repeatedly many of the thorny issues.

Finally, I wish to acknowledge with deep appreciation support from the Louis W. and Maud Hill Family Foundation and from the Carnegie Corporation.

H. F.

April 1967

Table of Contents

THE "MENTAL" AND THE "PHYSICAL"

The "Mental" and the "Physical"

I. A Preliminary Survey of Some Perplexities and Their Repression

Tough-minded scientists tend to relegate the mind-body problem to the limbo of speculative metaphysics. Perhaps after trying a bit but with questionable success to square themselves with the puzzle, they usually take one or the other of two attitudes. Either the puzzle is left to the philosophers to worry about, or else it is bluntly declared a pseudoproblem not worth pondering by anybody. Yet, the perplexities crop up again and again, often quite unexpectedly, if not in central issues of substantive scientific research, then certainly and at least in connection with the attempts to formulate adequately and consistently the problems, the results, and the programs of scientific inquiry. The disputes regarding the very subject matter and definition of psychology furnish a poignant illustration. Is it *mental experience* or is it *behavior*?

The behaviorist revolution in psychology, as well as its opposite philosophical counterpart, the phenomenalistic point of view in epistemology, each in its way, tried to obviate the problem. But all sorts of perplexities keep bedeviling both parties. The problem may be repressed, but repression produces symptoms, logical symptoms such as paradoxes or inconsistencies in this case. The behaviorist psychologist assimilates his method to that of the "objective" natural sciences. Scientific psychology, as the well known saying goes, having first lost its soul, later its consciousness, seems finally to lose its mind altogether. Behaviorism, now after more than forty years of development, shows of course many signs of mitigation of its originally rather harsh and radical position. It has availed itself of various clothings from the storehouse of philosophical garments. But despite the considerably greater scientific and logical sophistication in recent treatments of the issue, it is somewhat depressing to note that the main philosophical positions still are these: materialism, mentalism, mind-body interactionism, evolutionary emergence theories, psychoneurophysiological parallelism (epiphenomenalism, isomorphism, double aspect theories), and neutral monism. Characteristically, the phenomenalist and the behaviorist positions, refined descendants or variants respectively of the mentalistic and the materialistic philosophies, have been most forcefully advocated by the positivists of the last and of the present century. Positivism, more distinctly than any other point of view, with its notorious phobia of metaphysical problems and its marked tendency toward reductionism, was always ready to diagnose the mind-body puzzle as a *Scheinproblem*. Small wonder then that phenomenalism (or neutral monism) on the one hand, and physicalism on the other, have been the favored positions in various phases of the history of the positivistic outlook.

In the philosophy of the enlightenment of the eighteenth century we find the outspoken and clear-headed phenomenalism of Hume, but also the equally explicit, though more "simpliste" French materialism, especially of Baron d'Holbach. The German positivists of the nineteenth century, led by Mach and Avenarius, were essentially Humeans. And so was Bertrand Russell in one of the earlier phases of his epistemological odyssey. It was the combined influence of Russell's phenomenalism (or neutral monism) and of the logic of *Principia Mathematica* which led Carnap in his early work *Der Logische Aufbau der Welt* (1928) to elaborate in considerable detail and with remarkable precision a logical reconstruction of the relation between psychological and physical concepts. He chose as a basis for this reconstruction a set of neutral experiential data and showed how the concepts of various scientific disciplines can be constituted as logical constructions erected on a basis of concepts which refer to elements and relations of that (subjectless) raw material of immediate experience. Carnap's attempt was thus a culminating point in the series of positivistic-phenomenalistic epistemologies. But certain grave objections and difficulties soon made Carnap abandon this scheme and replace it by another, different in basis and structure. His new reconstruction is physicalistic in that the basic elements and relations are the designata of an intersubjective observation language (viz., the physicalistic thing-language). The difference in logical structure is due mainly to the recognition that the Russellian hierarchy of types does not adequately explicate the category mistakes which undoubtedly give rise to some (though by no means all) mind-body puzzles.

The physicalist views of Lashley (192), Carnap (62, 64, 66, 67), Hempel (146), Black (37, 38), Quine (268), Ryle (294), Skinner (321), and Wilfrid Sellars (315), though differing in many more or less important respects one from another, are primarily motivated by a basic doubt about the possibility of a purely phenomenal language. The observation language of everyday life, we are told, is rooted in the intersubjective terms whose usage we acquired in the learning situations of a common, public context of labeling things, properties, relations, states, events, processes, and dispositions. Subjective or "mentalistic" terms, this group of thinkers claims, are introduced and their usage learned on an intersubjective basis. Remove this intersubjective basis and not only have you deprived psychological concepts of their scientific significance, but you are left with nothing more than ineffable raw feels or with exclamations devoid of cognitive significance.

But the problems will not completely yield to this reductive approach. Introspection, though admittedly often unreliable, does enable us to describe elements, aspects, and configurations in the phenomenal fields of direct experience. When the doctor asks me whether I have a pain in my chest, whether my mood is gloomy, or whether I can read the fine print, *he* can afford to be a behaviorist and test for these various experiences in a perfectly objective manner. But *I have* (or do not have) the pain, the depressed mood, or the visual sensations; and I can report them on the basis of direct experience and introspection. Thus the question arises inevitably: how are the raw feels related to behavioral (or neurophysiological) states? Or, if we prefer the formal mode of speech to the material mode, what are the *logical* relations of raw-feel-talk (phenomenal terms, if not phenomenal language) to the terms and statements in the language of behavior (or of neurophysiology)?

No matter how sophisticated we may be in logical analysis or epistemology, the old perplexities center precisely around this point and they will not down. Many philosophical positions at least since the eighteenth century were primarily motivated, I strongly suspect, by the wish to avoid the mind-body problem. Moreover, the central significance of the problem for any *Weltanschauung* burdens its clarification with powerful emotions, be they engendered by materialistic, idealistic or theological prepossessions. Schopenhauer rightly viewed the mind-body problem as the "*Weltknoten*" (world knot). It is truly a cluster of intricate puzzles—some scientific, some epistemological, some syntactical, some semantical, and some pragmatic. Closely related to these are the equally sensitive and controversial issues regarding teleology, purpose, intentionality, and free will.

I am convinced, along with many contemporary philosophical analysts and logicians of science, that *all* of these problems have been unnecessarily complicated by conceptual confusions, and to that extent are gratuitous puzzles and pseudoproblems. But I feel that we have not yet done *full* justice to any of them. Repression by reductionist philosophies (positivism, phenomenalism, logical behaviorism, operationism) is fortunately going out of fashion and is being replaced by much more detailed and painstaking analyses, of both the (Wittgensteinian) "ordinary language" and the (Carnapian) "reconstructionist" types.

Collingwood once said "people are apt to be ticklish in their absolute presuppositions; [they] blow up right in your face, because you have put your finger on one of their absolute presuppositions." One might add that philosophers are hypersensitive also in their repressed perplexities. A puzzle which does not resolve itself within a given favored philosophical frame is repressed very much in the manner in which unresolved intrapersonal conflicts are repressed. I surmise that psychologically the first kind may be subsumed under the second. Scholars cathect certain ideas so strongly and their outlook becomes so ego involved that they erect elaborate barricades of defenses, merely to protect their pet ideas from the blows (or the slower corrosive effects) of criticism. No one can be sure that he is not doing this sort of thing in a particular case, and I claim no exception for myself. The best one can do is to proceed with candor and to subject oneself to ruthless criticism as often as feasible and fruitful. Techniques of self-scrutiny are nothing new in philosophy, but implemented by modern depth-psychological tools they could surely be made much more effective. In this candid spirit, I shall begin by putting my cards quite openly on the table; in the next two sections I shall indicate what I consider the sort of requirements for an adequate solution of the mind-body problems. I have no doubt whatever that some philosophers or psychologists will differ from me even in these first stages. All I can do then is to try, first to make these requirements as plausible as I can, and second, to analyze and evaluate the assets and the liabilities of some of the various proposed solutions as fully as space permits.

II. The Scientific and the Philosophical Strands in the Mind-Body Tangle

A first indispensable step toward a clarification of the issues is to separate the scientific from the epistemological questions pertaining to the relations of the mental to the physical. Epistemology is here understood in the modern sense of a logical analysis of concepts and statements and of the closely related logical reconstruction of the validation of knowledge claims. Some of the pertinent statements themselves are, however, essentially of a scientific nature in that they fall under the jurisdiction of empirical evidence. It is right here where we find a fundamental parting of the ways. Biologists, psychologists (and with them, many philosophers) hold deep convictions, one way or another, on the autonomy or non-autonomy of the mental. The strongest contrast is to be found between those who hold interactionistic views regarding the mental and the physical, and those who reject interactionism and hence espouse either parallelism (e.g., in its currently favored form, isomorphism) or some emphatically monistic view. Interactionism as well as parallelism is of course a form of dualism. The main difference and dispute between these two points of view is at present not fully decided by the evidence. But I think this is an issue to which empirical evidence is ultimately and in principle relevant.

Vitalists or interactionists like Driesch, McDougall, J. B. Pratt, Ducasse, Kapp, *et al.* hold that biological concepts and laws are not reducible to the laws of physics, and hence—a fortiori—that psychological concepts and laws are likewise irreducible. Usually this doctrine is combined with a theory of the emergent novelty of life and mind. But there are others who restrict emergence to the mental, i.e. they hold a reducibility view in regard to the biological facts. "Reducibility" is here understood to mean the same as "explainability"; and has no necessary connection with the introducibility (empirical anchorage) of biological or psychological concepts on the basis of physicalistic *observation* terms. As Carnap (67) has pointed out clearly, the thesis (*his* thesis) of the unity of the *language* of science does not in any way prejudge the issue of the unitary explainability of biological and psychological facts (or laws) on the basis of physical *theory*. Philosophers should certainly not assume that such a basic scientific issue can be settled *merely* by logical analysis. It is *logically* conceivable that biological, psychological, and social phenomena (as well as their regularities) may not be explainable in terms of those physical or physicochemical laws (and theoretical assumptions) which are sufficient for the explanation and prediction of inorganic phenomena (and their regularities).

Logical parallels to such irreducibilities are clearly evident even *within* physics. The "mechanistic" (Newtonian) premises of explanation are now viewed as entirely insufficient for the explanation of electromagnetic radiation, of the dynamics of intra-molecular and intra-atomic processes, and of the interaction of electromagnetic radiation and the particles of matter. Nineteenth century physics added the fundamentally new concepts and laws of electromagnetics; and these in turn were drastically modified and supplemented by the relativity and quantum theories of our century. It is conceivable that homologous emendations may be required for the explanation of the phenomena of life and mind. Contemporary dualists, be they vitalists, emergentists, interactionists, or parallelists, maintain that such an enrichment of the conceptual system of science will be indispensable. Their arguments are based primarily on the traditionally captivating evidence of teleological processes, purposive behavior, psychosomatics, and the mnemonic and intentional features of perception, cognition, thought, desire, and volition. And some apparently very persuasive arguments point simply to the existence (occurrence) of immediate experience, i.e., the raw feels or hard data of the directly given. They maintain that these data, though *related* to behavior and neurophysiological processes, are not *reducible* to, or *definable* in terms of, purely physical concepts; and that their occurrence is not predictable or explainable on the basis of physical laws and physical descriptions only.

At this point the distinction between the scientific and the philosophical aspects of the mind-body problems becomes imperative. "Irreducibility" may mean non-derivability from a specified set of premises; but in other contexts it may mean non-translatability (non-synonymy, non-equivalence in the *logical* sense). To illustrate: many physical phenomena of sound or heat are derivable from the kinetic theory of molecular motion. In this sense certain parts of acoustics and of thermodynamics are reducible to mechanics, with a high degree of approximation at least within a certain limited range of the relevant variables.

But the phenomena of heat radiation (and similarly those of optics, electricity, magnetism, and chemistry) are not reducible to mechanics. Whitehead speculatively maintained that the laws pertaining to the motion of electrons in living organisms differ fundamentally from the laws of electrons in the context of inorganic lifeless bodies. In a similar vein the physicist Elsasser (95, 96, 97), following some suggestions contained in Bergson's views on organic life and memory, regards the physical laws as special or limiting cases of biological laws. This is a drastic reversal of the "Victorian" outlook according to which macro-regularities are (usually) explainable in terms of basic micro-laws.*

As a student of the history and the methodology of modern science, and impressed as I am with the recent advances of biophysics, biochemistry, and neurophysiology, I am inclined to believe strongly in the fruitfulness of the physicalistic research program (involving micro-explanations) for biology and psychology. But qua analytic philosopher my intellectual conscience demands that I do not prejudge the issues of reducibility (explainability) in an a priori manner. Beyond the sketchy *empirically* oriented arguments which I am going to submit presently, I shall address myself later on primarily to the logical and epistemological aspects of the mind-body problem.

Along empirical lines I believe there are differences, in principle capable of test, between parallelism and interactionism (and/or emergentism). Psycho-neurophysiological parallelism is here understood as postulating a one-one, or at least a one-many, simultaneity-correspondence between the mental and the physical. Parallelism as customarily conceived clearly rules out a many-one or a many-many correspondence. This latter type of correspondence, if I may speak for a moment about the *motivation* rather than the evidential substantiation (confirmation), is generally unpalatable to the scientific (especially the "Victorian") point of view, because it would obviously limit the predictability of mental events from neurophysiological states of the organism. But given a "dictionary," i.e., more properly speaking, a set of laws correlating in one-one or many-one fashion physical and mental states, physical determinism is not abrogated.

* I have dealt elsewhere (106, 108, 112, 113, 115, 116) with the logic and methodology of such explanations. See also the important articles by E. Nagel (230, 232); Hempel and Oppenheim (152); Kemeny and Oppenheim (177); Oppenheim and Putnam (in *Minnesota Studies in the Philosophy of Science*, Vol. II).

Two important qualifying remarks are in order here: (1) By "physical determinism" I mean, of course, that degree of precise and specific in-principle-predictability that even modern quantum physics would allow as regards the macro- and some of the micro-processes in organisms. (2) By "physical" I mean * the type of concepts and laws which suffice in principle for the explanation and prediction of inorganic processes. If emergentism is *not* required for the phenomena of organic life, "physical" would mean those concepts and laws sufficient for the explanation of inorganic as well as of biological phenomena. In accordance with the terminology of Meehl and Sellars (221), I shall henceforth designate *this* concept by "*physical*$_2$" in contradistinction to "*physical*$_1$", which is practically synonymous with "scientific", i.e., with being an essential part of the coherent and adequate descriptive and explanatory account of the spatio-temporal-causal world.

In view of what was said above about the *empirical* character of the interaction and the emergence problems, the concepts of mental states might well be *physical*$_1$ concepts, in that they could be introduced on the basis of the intersubjective observation language of common life (and this includes the observation language of science). Just as the concept of the magnetic field, while not denoting anything directly observable, can be introduced with the help of postulates and correspondence rules (cf. Carnap, 73), so it is conceivable that concepts of vital forces, entelechies, "diathetes" (cf. Kapp, 172, 173, 174), and mental events might be given their respective meanings by postulates and correspondence rules. Of course, the question remains whether such ("emergent") concepts are really needed and whether they will do the expected job in the explanation and prediction of the behavior of organisms, subhuman or human. My personal view, admittedly tentative and based on the progress and partial success of physicalistic microexplanation (implemented by Gestalt and cybernetic considerations), is to the effect that *physical*$_2$ laws will prove sufficient. But, having abandoned the all too narrow old meaning criteria of the earlier logical positivists, I would not for a moment wish to suggest that the doctrines of emergence or of interactionism are scientifically meaningless.

Let us then return to the empirically testable difference between interactionism and emergentism on the one hand, and parallelism on

*** In *this* context only; other meanings of "physical" will be listed and discussed in sections IV and V.**

the other. An obvious and picturesque analogy or model for the interactionist view may be suggested here to provide a more vivid background. Billiard balls are in motion on a billiard table, and their motions are, we assume, predictable on the basis of mechanical laws (Newton's, supplemented by the laws of friction and of partially elastic collision). But imagine now a mischievous boy standing by, once and again pushing this or that ball or lifting some ball from the table. The mechanical laws, combined with a statement of initial conditions for the balls and the table, at a given moment, will then no longer suffice for the prediction of the course of the balls. The system in this case is of course an open one. If we could proceed to a larger closed system including the boy, with information about his shifting desires and so forth, deterministic predictability might be restored. (Since prediction of the boy's actions is precisely the issue at stake, I shall not beg any questions here and shall leave the boy's behavior unexamined for the moment.) This model is merely to illustrate a good clear meaning of "interaction". The boy watches the balls and his actions are in part influenced by their momentary distribution and motions on the board. The events on the board are in turn influenced by the boy's actions. From the point of view of ordinary usage, it is proper to employ the word "interaction" perhaps only when we deal with causal relations directed both ways between two continuants (things, organisms, persons, etc.).

But even a theory of emergence, such as the one suggested, though not definitely endorsed, by Meehl and Sellars (221), is confirmable in principle by showing that $physical_2$ determinism does not hold. Mental states or raw feels, be they regarded as states of an interacting substantial mind (or soul) or as values of emergent scientific variables, would in any case entail a breach in *physical*$_2$ determinism. The system of neurophysiological events inasmuch as it is describable in $physical_2$ terms would have to be regarded as open not only in the usual way, i.e., in regard to the extraneural, let alone extradermal, events, but it would also be open in regard to the set of mental events with which they are assumed to be causally (functionally) related in a way that would make them radically different from a set of mere epiphenomena. Now, while it is admittedly difficult at present to test for the implied breach in *physical*$_2$ determinism, the idea is not metaphysical in the objectionable sense that empirical evidence could not conceivably confirm or disconfirm it.

Much depends in this issue upon just how the "interactors" or the "emergents" are conceived. Traditional vitalism, culturally and historically perhaps a descendant of more primitive forms of animism, stresses the *capricious* nature of *vis vitalis* and of *anima*. (In our model the boy by the billiard table is assumed to exercise "free" choice.) But interaction need not be indeterministic in the wider system. The wind and the waves of the sea genuinely interact; even if the wind's influence is quantitatively greater, the waves do have some effect upon the air currents nearby. But though precise prediction of detail is practically extremely difficult because of the enormous complexities of the situation, this type of interaction is in principle deterministically * analyzable in terms of the functional relations of the two types of variables. Even the individual "free" or "capricious" momentary choices of our boy might be predictable in principle; but here the practical feasibility is far beyond the horizon of current psychology. At best only some statistical regularities might be formulated.

Determinism, inasmuch as it is allowed for by current physical theory, is also the presupposition of the sophisticated conception of emergence as presented in the essay by Meehl and Sellars. Here we have no interacting things or substances, but scientific variables intertwined in such a way that certain values in the range of one set of variables are functionally so related to the values of the variables in the other set, that the relations in the second set are nomologically different from what they would be if the values of the first set are zero. More concretely, once mental states have emerged, their very occurrence is supposed to alter the functional relations between the neurophysiological ($physical_2$) variables in a manner in principle susceptible to confirmation. While my (scientific) predilections are completely incompatible with this ingenious and fanciful assumption, I do consider it scientifically meaningful. I just place my bets regarding the future of psychophysiology in the "Victorian" direction. And I admit I may be woefully wrong.†

* Again it is only to the extent that hydrodynamics and aerodynamics for macroprocesses are (approximately) deterministic.

† In his earlier formulations of the general theory of relativity Einstein endorsed the so-called Mach principle, according to which centrifugal and inertial forces are the effects of accelerations relative to the total masses of the fixed-stars-universe. But, impelled by what he considered cogent physical and mathematical arguments, he later ascribed those effects to a relatively independently existing "Führungsfeld" (guiding field). I mention this merely as a somewhat remote logically parallel case from an entirely different domain of science. Naturally, my expectation here is that something

With the foregoing remarks I hope to have indicated clearly enough that I consider these basic issues as essentially scientific rather than philosophical. But a full clarification and analysis of the precise meanings and implications of, respectively, parallelism, isomorphism, interactionism, and the various forms, naive or sophisticated, of emergentism is a *philosophical* task. I shall now develop the philosophical explication of the factual-empirical meaning of these assorted doctrines a little further and bring out their salient epistemological points. Parallelism and isomorphism, now that we have recovered from the excesses of positivism and behaviorism, are generally considered as inductively confirmable hypotheses. Reserving more penetrating epistemological analyses, especially of the "immediate experience" and "other minds" problems, until later, I assume for the present purpose and in the vein of the recent positions of Ayer (15, 18) and Pap (243, 248) that the ψ-Φ (i.e., psycho-neurophysiological) relations or correspondences can be empirically investigated; and that mental states (raw feels) may by analogy be ascribed to other human beings (and higher animals), even if in the case of those "others" they are inaccessible to *direct* confirmation.

Parallelism, then, in its strongest form assumes a one-to-one correspondence of the ψ's to the Φ's. It is empirically extremely likely that these correspondences are *not* "atomistic" in the sense that there is a separate law of correspondence between each discernible ψ_1 and its correlate Φ_1. It is quite plausible that, for example, different intensities of a phenomenally given tone (e.g., middle C), at least within a given range, are correlated with corresponding values in a limited range of some variable(s) of the neural processes in the temporal lobe of the brain.

Isomorphism as understood by the Gestalt psychologists (Wertheimer, Köhler, and Koffka) and the cyberneticists (Wiener, McCulloch, Pitts, etc.) assumes an even more complete one-one correspondence between the elements, relations, and configurations of the phenomenal fields and their counterparts in the neurophysiological fields which characterize portions of cerebral, and especially cortical, processes. As mentioned before, this sort of approach would also countenance a one-many correspondence of ψ's and Φ's. In that case, mental states would

of Mach's principle, even if in strongly modified form, will be salvaged. Powerful inertial forces as effects of a self-existent metrical field seem extremely implausible to me.

(with the help of the ψ-Φ "dictionary") still be uniquely inferable from neurophysiological descriptions. But many-one or many-many correspondences, even if expressed in terms of statistical laws, would seriously restrict such inferences from specific Φ's to specific ψ's. I know of no good empirical reasons for assuming anything but one-one correspondence; or one-many if very exact and detailed Φ-descriptions are used, and if account is to be taken of the limited introspective discernibility of the ψ's from one another.

Interactionism, as I understand (but reject) it, would entail a many-one or many-many correspondence. Arthur Pap (242, p. 277), however, argued that there is no empirically confirmable difference between parallelism and interactionism. This, he thought, is because lawful relations or functional dependencies are the modern scientific equivalent of the cause-effect relation. Temporal succession, he maintains, is not a criterion of causal connection. While I admit that the most general conception of the causal relation is simply that of a (synthetic) sufficient condition,* and is thus free of any connotation regarding the temporal succession of cause and effect; and though I also agree that in the case of ψ-Φ relation it would seem rather fantastic to assume anything like a time difference, I think that the interaction hypothesis differs in its empirical meaning from parallelism or isomorphism in that it entails a breach of physical$_2$ determinism for the Φ's. This, if true, could in principle be confirmed by autocerebroscopic evidence. For example, the experience of volitions as directly introspected would not be correlated in one-to-one (or one-many) fashion with simultaneous cortical states as observed (really inferred) by looking upon the screen of a cerebroscope,† and regularly succeeded by certain processes in the efferent nerves of the brain, ultimately affecting my muscles or glands, and thus ensuing in some act of behavior. This is the sort of most direct evidence one could ever hope for, as regards the confirmation of ψ-Φ action. If the idea of *interaction*, i.e., action both ways between the ψ's and the Φ's, is entertained, then there should be sensations (produced by the

* And in the laws of classical mechanics and electrodynamics of sufficient and necessary condition.

† This, for the time being, of course, must remain a piece of science fiction (conceived in analogy to the doctors' fluoroscope) with the help of which I would be able to ascertain the detailed configurations of my cortical nerve currents while introspectively noting other direct experiences, such as the auditory experiences of music, or my thoughts, emotions, or desires.

chain of processes usually assumed in the causal theory of perception, but) not strictly correlated with the terminal cortical events.

Characteristically, philosophers have been emphasizing much more the action of "mind on matter"—as in voluntary behavior, or in the roles of pleasure, pain, and attention—than that of "matter on mind." This asymmetrical attitude usually comes from preoccupation with the freewill puzzle, or related to this, from some remnants of theological ideas in the doctrines of an ideal ("noumenal") self. But the freewill puzzle—even if some details of its *moral* aspects still await more clarification—has in its scientific aspects been satisfactorily resolved by making the indispensable distinctions between causality and compulsion (and indeterminism and free choice). The perennial confusions underlying the freewill perplexity, truly a scandal in philosophy, have been brilliantly exposed by empiricist philosophers.*

The main reasons why most psychophysiologists (and along with them many philosophers) reject the hypothesis of ψ-Φ-many-one or many-many correspondence are these:

1. Normal inductive extrapolation from the successes of psychophysiology to date makes it plausible that an adequate theory of animal and human behavior can be provided on a neurophysiological basis. Most physiologists therefore favor ψ-Φ parallelism or epiphenomenalism. Parallelism, I repeat, is here understood as the assertion of the one-one (or, at least, one-many) ψ-Φ correspondence, and not, as by Wundt and some philosophers, as the doctrine of double causation, i.e., involving parallel series of events with temporal-causal relations corresponding (contemporaneously) to one another on both sides. Causality in the mental series is by far too spotty to constitute a "chain" of events sufficiently regular to be deterministic by itself. Epiphenomenalism in a value-neutral scientific sense may be understood as the hypothesis of a one-one correlation of ψ's to (some, not all) Φ's, with determinism (or as much of it as allowed for by modern physics) holding for the Φ-series, and of course the "dangling" nomological relations connecting the Φ's with the ψ's. According to this conception voluntary action as well as psychosomatic processes, such as hysteria, neurotic symptoms,

* Hobbes, Locke, and especially Hume, Mill, Sidgwick, Russell, Schlick (301); and Dickinson S. Miller, cf. the superb article he published under the pseudonym "R. E. Hobart" (157). See also C. L. Stevenson (329); University of California Associates (339); A. K. Stout (330); and Francis Raab (271).

and psychogenic organic diseases (e.g., gastric ulcers) may ultimately quite plausibly be explained by the causal effects of cerebral states and processes upon various other parts of the organism; only the cerebral states themselves being correlated with conscious (or unconscious *) mental states.

2. While the cultural and historical roots of the epiphenomenalist doctrine may be the same as those of traditional materialism, we can disentangle what is methodologically sound and fruitful in the materialistic point of view from what is cognitively false, confused, or meaningless. The fundamental methodological reason for the rejection of interactionism, or the (equivalent) adoption of ψ-Φ-one-one (or one-many) correspondence as a working hypothesis or research program, however, is this: If the ψ's are not inferable on the basis of intersubjectively accessible (observed, or usually, inferred) Φ's, then their role is suspiciously like that of a *deus ex machina*. The German biologist-philosopher Driesch admitted this candidly, and thereby gave his case for vitalism away. He said that the intentions of the entelechy could be inferred only *post factum*, but could not be predicted from antecedent physical conditions. This is just like the case, in our crude analogy, of the capricious boy at the billiard table. After he has removed a ball we may say that he intended (perhaps!) to avoid a collision of the red ball with the white one. According to the vitalist interactionist doctrine, the volitions of the boy are in principle unpredictable on the basis of any and all antecedent conditions in his organism and the environment. Interactionism so conceived assumes causal relations between the elements in the series of mental states, the series of physical states, but also some crossing from the set of mental states to the physical ones and vice versa. In the model of the wind and the waves, we have precisely this sort of schema exemplified. But notice the crucial difference. A closed system (or a system with known initial and boundary conditions) is here conceivable in which all relevant variables are ascertainable *intersubjectively* and *antecedently* to the prediction of later states of the system; whereas in the case of ψ-Φ interaction, intersubjective and antecedent confirmation of the ψ-states is ex *hypothesi* excluded.

The flavor of the theological arguments from design and of primitive animistic explanations of nature and human behavior permeates inter-

* The terminological question whether to speak of the unconscious as "mental" will be discussed in sections IV and V.

actionistic explanations. They are at best *ex post facto* explanations. This sort of explanation, while not as satisfactory as explanations that also have *predictive* power, is nevertheless quite legitimate and is frequently the best we can provide in complex situations. Earthquakes are notoriously unpredictable (i.e., *practically* unpredictable), but once we observe a certain case of large scale destruction, its explanation in terms of an earthquake is perfectly legitimate even if the precise location of each piece of rubble in the shambles is far from predictable. Biologists are satisfied with evolutionary (retrospective) explanations of the emergence of a new species, even though they could never have predicted this emergence in any specific detail. Given the species in the Cambrian epoch, and given the principles of genetics and of Neo-Darwinian evolution, nobody could inductively infer the emergence of the chimpanzee or of the orchid; nevertheless, the very partial explanations of the theory of evolution are scientifically significant, acceptable, and helpful. Explanations of historical phenomena like wars, revolutions, and new forms of art furnish another illustration for the same type of *ex post facto* explanations. Finally, for an example in the psychological domain, if we find that a man has written dozens of letters of application for a certain type of job, we infer that he was impelled by a desire for such a job, even if we could not have predicted the occurrence of this desire on the basis of antecedent and intersubjectively confirmable conditions.

It is important, however, to notice again the decisive difference between explanations for which it is at least in principle conceivable that they could be predictive (as well as retrodictive), and those which *ex hypothesi* are *only* retrodictive. Scientists are predominantly interested in enlarging the scope of predictive explanations. The opposition against vitalism then stems from a reluctance to admit defeat as regards predictability. And the opposition against ψ-Φ interactionism stems furthermore from the reluctance to admit antecedents which are only subjectively accessible into the premises (regarding initial conditions) for predictive inferences. Expressing the same idea positively, we may say that it is part of the methodology or of the over-all working hypothesis of modern science that prediction, to the extent that it is possible at all (taking account of the basic quantum indeterminacies), is always in principle possible starting from *intersubjectively* confirmable statements about initial conditions. Scientists have, on the whole, adjusted themselves to the limitations involved in *statistical* prediction and *probabilistic* explanation. Very likely nothing better will ever be forthcoming in any area except in the few where classical determinism holds with a high degree of approximation. Of course, a logical distinction should be made between those cases in which the restriction to probabilistic predictability is a consequence of the complexity of the situation, and those in which the *theoretical* postulates of a given domain are themselves formulations of statistical laws. Although one can never be sure that this distinction is correctly drawn or that the dividing line will remain in the same place during the progress of science, the distinction *can* be drawn tentatively in the light of theories well confirmed at a given time.

But scientists are radically opposed to the admission of *purely subjective* factors or data (conceived as in principle inaccessible to intersubjective confirmation) as a basis for prediction or explanation. This would indeed be scientifically meaningless, if not even statistical relations of subjective states to antecedent or consequent intersubjective observables could be assumed. If they *are* assumed, then the subjective states are not *purely* subjective or "private" in the radical sense intended by some interactionists. The "emergent" raw feels in the interpretation by Meehl and Sellars are of course subjective only in the sense that they can be the objects of direct introspective verification, but they are also intersubjective (physical_1) in the sense that they can be assumed (posited, inferred, hypothetically constructed) by scientists who do not have the same sort of raw feels in the repertory of their own direct experience. This is so, for example, in the case of a congenitally blind scientist equipped with modern electronic instruments who could establish the (behavioristic) psychology of vision for subjects endowed with eyesight. The blind scientist could thus confirm all sorts of statements about visual sensations and qualities—which in his knowledge would be represented by "hypothetical constructs." But if *ex hypothesi* all connections of the subjective raw feels with the intersubjectively accessible facts are radically severed, then such raw feels are, I should say by definition, excluded from the scope of science. The question whether discourse about such absolutely private raw feels makes sense in any sense of "sense" will be discussed later.

The upshot of this longish discussion on the difference between the scientific and the philosophical components of the mind-body problems is this: If interactionism or any genuine emergence hypotheses are sensibly formulated, they have empirical content and entail incisive limitations of the scope of $physical_2$ determinism. Interactionism is more difficult to formulate sensibly than is the (Meehl-Sellars) emergence hypothesis. In one form it requires substances (things, continuants or systems of such) for a normal use of the term "interaction," and in this form there seems little scientific evidence that would support it. I have read a great many arguments by metaphysicians attempting to support the idea of a totally (or partially) immaterial "self." But I have never been able to discern any good *cognitive* reasons beneath their emotionally and pictorially highly charged phrases. Whatever role the self (in Freudian terms perhaps the total superego, ego, and id-structure) may play in the determination of human conduct, it may yet very well be explained by a more or less stable structure of dispositions due to some constitutionally inherited, maturationally and environmentally modified, and continually modulated structure of the organism (especially the nervous and endocrine systems).

In another form interactionism (without a self) would require "spontaneously" arising mental states, i.e., an indeterminism not even limited by statistical regularities, and this again is neither supported by empirical evidence, nor advisable as a regulative idea for research. Nor is it required for the solution of the freewill problem, or for an account of the causal efficacy of mental events in the course of behavior. As regards the emergence hypothesis (à la Meehl and Sellars), this clearly makes sense, but whether it is really needed for the explanation of behavior is an open question. In the spirit of the normal procedures of scientific induction and theory construction I remain conservative in thinking that the rule of parsimony (Ockham's razor, or Newton's first *regula philosophandi*) warns us not to multiply entities (factors, variables) beyond necessity. If the necessity should become evident in the progress of research, I shall cheerfully accept this enrichment of the conceptual apparatus of science; or, ontologically speaking, this discovery of new entities in our world. In the meantime, I remain skeptical about emergence, i.e., optimistic about the prospects of $physical_2$ determinism. And, as I shall argue from the point of view of epistemology in sections IV and V, the sheer existence of raw feels is not a good reason for holding an emergence doctrine.

Another philosophical issue which needs careful separation from the scientific problems among the mind-body tangles is that of the "intentionality" of the mental. (For expository reasons the discussion of this issue will be reserved for section IV *F*.)

III. Requirements and Desiderata for an Adequate Solution of the Mind-Body Problem. A Concise Statement of the Major Issues

If the title of this section were not already a bit too long, I should have added, "as *I* view these requirements and desiderata, and as *I* conceive the adequacy of a solution." All I can say by way of extenuation of my personal biases in this matter is that I have concerned myself seriously and repeatedly with the problem for about thirty-six years; that I have studied most of the contributions from thinkers of many lands in modern and recent philosophy and science; and that this is my fourth published attempt to arrive at an all around satisfactory clarification. There have often been moments of despair when I tried ineffectively to do justice to the many (apparently) conflicting but impressive claims coming from ever so many quarters. It is, then, with a heavy sense of intellectual responsibility and not without some misgivings that I proceed to enumerate the following requirements, desiderata, and considerations which seem to me the conditions (or at least *some* of the conditions) that may serve as criteria of adequacy for a solution of the problem; a solution that is to be satisfactory from the point of view of contemporary science as well as in the light of modern philosophical analysis. I concede unblushingly that in some respects I share here the attitude of some of the (shall I say, epistemologically not too naive) metaphysicians who have wrestled with the problem and have tried to provide a solution that is *synoptic* in that it would render a just, consistent, and coherent account of all relevant aspects and facets of the issue.

Here, then, is my list of requirements and desiderata (or "conservanda" and "explicanda"):

1. The terms "mental" and "physical" are precariously ambiguous and vague. Hence a first prerequisite for the clarification and the adequate settlement of the main issues is an analytical study of the meanings of each of these two key terms, and a comparative critical appraisal of the merits and demerits of their various definitions and connotations. Due attention will also be given to the (partly) terminological question as to whether to include under "mental" beside the directly experienced

and introspectible also the unconscious states and processes of depth-psychological theories (Freudian or Neo-Freudian). All this will be undertaken in the next section of this essay.

2. In the light of what was said in the preceding section about the scientific (empirical) components of the mind-body problem, an analysis of the mind-body relation is to be sought which does justice to the arguments for the sort of mind-body unity which impresses itself increasingly upon the majority of psychologists, psychophysiologists, and psychiatrists of our time. Although the question of evolutionary as well as of logical "emergence" cannot be decided by a priori philosophical considerations, vitalistic and interactionist doctrines appear on empirical and methodological grounds as suspect and undesirable. Just what the alleged facts of parapsychology (telepathy, clairvoyance, precognition, psychokinesis, etc.) may imply for the mind-body problem is still quite unclear. Here too, it seems to me, any speculations along the lines of interactionism are—to put it mildly—premature, and any theological interpretations amount to jumping to completely unwarranted conclusions. My own attitude in regard to the experiments (statistical designs) on extrasensory perception, etc. is that of the "open mind." The book by Soal and Bateman (325) and its discussion by M. Scriven (305) present evidence and arguments which can not lazily or cavalierly be shrugged off. The chances of explaining the "facts" away as due to experimental or statistical error, let alone as outright hoax or fraud, seem now rather remote. But even granting these facts, I think that efforts should be made to explain them first by revisions and emendations in the *physical* theory of behavior before we indulge in speculations about immaterial souls or selves. These remarks clearly reveal my bias in favor of a naturalistic, if not monistic, position. That and how this position differs from "crass materialism," the bugbear of idealistic and spiritualistic metaphysicians, will be explained later on.

3. Any solution of the mind-body problem worth consideration should render an adequate account of the efficacy of mental states, events, and processes in the behavior of human (and also some subhuman) organisms. It is not tendermindedness or metaphysical confusions, I trust, which impel this repudiation of a materialistically oriented epiphenomenalism. Admittedly, the testimony of direct experience and of introspection is fallible. But to maintain that planning, deliberation, preference, choice, volition, pleasure, pain, displeasure, love, hatred, at-

tention, vigilance, enthusiasm, grief, indignation, expectations, remembrances, hopes, wishes, etc. are not among the causal factors which determine human behavior, is to fly in the face of the commonest of evidence, or else to deviate in a strange and unjustifiable way from the ordinary use of language. The task is neither to repudiate these obvious facts, nor to rule out this manner of describing them. The task is rather to analyze the logical status of this sort of description in its relation to behavioral and/or neurophysiological descriptions. In the pursuit of this objective it will of course be necessary to avoid both interactionism and epiphenomenalism; and it will moreover be desirable to formulate the solution in such a way that it does not presuppose emergentism (in the sense of $physical_2$ indeterminism), although the door to a scientifically formulated emergentism need not be closed.

In this same connection justice should be rendered to what is meaningful and scientifically defensible in the notion of free will or free choice. If our personality-as-it-is at the moment of choice expresses itself in the choice made; if our choices accord with our most deeply felt desires, i.e., if they are not imposed upon us by some sort of compulsion, coercion, or constraints such as by brute physical force, by other persons (or even only by components of our personality which we do not acknowledge as the "core" deemed centrally our "self"), then we are "free" in the sense that we are the doers of our deeds, the choosers of our choices, the makers of our decisions. In other words, it is in this case that our central personality structure is a link in the causal chain of our behavior, predominantly, even if not exclusively, effective in the determination of our conduct. This sort of freedom (in the superb formulation of R. E. Hobart-Dickinson Miller) "*involves determinism and [is] inconceivable without it.*" *

4. A most important *logical* requirement for the analysis of the mind-body problem is the recognition of the *synthetic* or *empirical* character of the statements regarding the correlation of psychological to neurophysiological states. It has been pointed out time and again † that the early reductionistic logical behaviorism failed to produce an adequate and plausible construal of mentalistic concepts by explicit definition on the basis of purely *behavioral* concepts. (In the less adequate material

* Cf. R. E. Hobart (157).

† Cf. F. Kaufmann (175), N. Jacobs (163), C. I. Lewis (196), E. Nagel (230), A. Pap (243), *et al.*

mode this might be put by bluntly saying that mind is not identifiable with behavior.) For a long time, however, I was tempted to identify, in the sense of *logical* identity, the mental with the neurophysiological, or rather with certain configurational aspects of the neural processes. It was in this sense that I (103) suggested a *double-language* theory of the mental and the physical. But if this theory is understood as holding a *logical translatability* (analytic transformability) of statements in the one language into statements in the other, this will certainly not do. Interlinguistic translations like "*Il ne fait pas beau temps*" into "The weather is not fine" are analytic if the respective meanings are fixed with the help of syntactical and semantical metalanguages common to both French and English. Similarly the geocentric description of the pure kinematics of the planetary system is analytically translatable into the corresponding heliocentric description, precisely because we avail ourselves here of transformation rules in a four-dimensional geometry (i.e., kinematics).

But the question which mental states correspond to which cerebral states is in some sense (to be analyzed epistemologically later on) an empirical question. If this were not so, the intriguing and very unfinished science of psychophysiology could be pursued and completed by purely a priori reasoning. Ancient and primitive people had a fair amount of informal and practical psychological knowledge, but the fact that mental states are closely associated with cerebral states was unknown to them. Aristotle held that the seat of our feelings and emotions is the heart (and this has survived in the traditions of poetic discourse). But to say that Aristotle was wrong means that we have now empirical evidence which proves that the emotions are linked to brain processes. It is therefore imperative to preserve the *synthetic* character of the assertion of this knowledge claim, whatever specifically may prove to be its most clarifying formulation.

If any of my readers should be hard-boiled behaviorists or "crass" materialists, it will be difficult to convince them that there is a problem at all. I can do no more than to ask them such persuasive or *ad hominem* questions as, Don't you want anesthesia if the surgeon is to operate on you? And if so, what you want prevented is the occurrence of the (very!) raw feels of pain, is it not? If you have genuine concern and compassion for your fellow human beings (as well as perhaps for your dogs, horses, etc.), what is it that you object to among the consequences of cruel treatments? Is it not the pains experienced by these "others"? It could not be merely their physical mutilation and consequent malfunctioning. Moral condemnation of wanton cruelty presupposes the meaningfulness of the ascription of direct experience to others. Subjective experience in *this* sense cannot be *logically* identical with states of the organism; i.e., phenomenal terms could not explicitly be defined on the basis of $physical_1$ or $physical_2$ terms.

It should be noted that we repudiate the *logical* translatability thesis not because of the possibility, definitely contemplated, of a one-many-ψ-Φ correspondence. One could always formulate such a correspondence with the help of a general equivalence between statements containing single ψ-predicates on the one side and disjunctions of statements containing several and various Φ-predicates on the other. It is rather the *logical necessity* of the equivalence which is here rejected. The equivalence must be construed as logically contingent.

5. Consonant with the spirit of the preceding discussions, but now to be stated explicitly, are three very closely related *epistemological* requirements. To list them first very briefly, they are:

(a) the need for a *criterion of scientific meaningfulness* based on *intersubjective* confirmability;

(b) the recognition that epistemology, in order to provide an adequate reconstruction of the confirmation of knowledge claims must employ the notion of immediate experience as a confirmation basis (the "given" cannot be entirely a myth!); "Acquaintance" and "Knowledge by Acquaintance," however, require careful scrutiny;

(c) the indispensability of a *realistic*, as contrasted with operationalistic or phenomenalistic, interpretation of empirical knowledge in general, and of scientific theories in particular.

(ad a) It is generally agreed that scientific knowledge claims must not only be intersubjectively communicable (intelligible), but also intersubjectively testable. The following considerations will illustrate the point. If the stream of my conscious experience continued beyond the death and decay of my body, then this may be verifiable by me (in some, none too clear, sense of "me"; but I shall let this pass for the moment). If such survival were, however, not even extremely indirectly or incompletely confirmable by others; if it were in no way lawfully connected with, and thus not inferable from, any feature of life (mine or that of others) before death, then, while the statement in question may be said to have subjective meaning, it could not become part of science in the sense in which "science" is commonly understood.*

(ad b) Recent behavioristic and physicalistic arguments to the contrary notwithstanding, I am still convinced that purely phenomenal statements make sense and are the ultimate epistemic basis of the confirmation (or disconfirmation) of knowledge claims. By this I do not at all wish to suggest that phenomenal statements are infallible ("incorrigible"), nor that they necessarily have a higher degree of certainty than intersubjectively confirmable statements about the ordinary objects of our common life environment. I grant that, especially for the purposes of the philosophy of *science*, it is more useful to choose the physicalistic thing language for the confirmation basis of knowledge claims. But when I judge, e.g., that a certain pain is increasing, or that I hear a certain ringing sound (no matter whether this sound-as-experienced is causally due to a doorbell, a police car siren, to "buzzing in my ear," or to a hallucination), then that certain *it* which may later find its place in the causal structure of the world is first of all, and taken by itself, a *datum of direct experience*. Whether I get to it "post-analytically," or whether I simply *have* it, pre-analytically; that is to say, whether I arrive at it by a kind of analysis starting from "seeming," "appearing," "looks like" ("sounds like," etc.) sentences; or whether I can by simultaneous introspection (self-observation) or immediate retrospection, ascertain the occurrence of a certain datum, I have no doubt that talk about phenomenal data and phenomenal fields makes sense; and that in a rational reconstruction of the confirmation of ordinary observation statements, we can (if we wish) penetrate to this deepest level of evidence.†

I have not been convinced by the arguments of Popper (258) that the search for "hard data" is doomed to failure, that the "given" is like a bottomless swamp. Nor am I convinced that a purely private language ‡ is inconceivable. Of course, if by "language" one means an in-

* For a fuller discussion of the scientific meaning criterion cf. my articles (103, 105, 109, 110, 114, 116) and Carnap (64, 67, 73). For stimulating discussions of the meaning of "disembodied minds" see Aldrich (6) and Lewy (199).

† For persuasive arguments along these lines, cf. B. Russell (284, 287); H. H. Price (264); C. I. Lewis (195, 197, 198); Ayer (12, 13, 18); N. Goodman (135, 136, 137). For an incisive critique of the "incorrigibility" arguments, cf. K. R. Popper (258); R. Carnap (62, 64); H. Reichenbach (273, 276); M. Black (38); J. Epstein (98).

‡ Cf. the symposium by Ayer and Rhees (16, 278).

strument of interpersonal communication, then the idea of an absolutely private language is self-contradictory. But, granting that in the normal case the capacity for using a language is acquired by education, it is not *logically* inconceivable that a child growing up in complete solitude might devise his own symbolism not only for the objects and events in its environment *but also* for the *raw feels* of its direct experience. Such a child might well come to use terms for various aches, pains, itches, tickles, moods, emotions, etc. I do not for a moment deny that the use of such subjective terms, in the usual and normal case, is acquired through trial and error learning, and in this process largely inculcated in the child by other persons who *tell* him, e.g., "now you are tired," "now you are glad," "you must have an awful pain." Such *tellings* by others are guided by the facial expressions, vocal emissions, posture, etc., i.e., generally by the observable behavior of the child (and by test condition → test result sequences in its behavior, involving both environmental stimulus situations and a variety of responses).*

In sum, I believe that there is an indispensable place for "acquaintance" and "knowledge by acquaintance" in a complete and adequate epistemology. A more detailed account and analysis of the meanings of these terms will be given in the two subsequent sections of the present essay.

(ad c) The last epistemological requirement, to be briefly discussed here, is that of a realistic, rather than phenomenalistic or operationalistic, reconstruction of knowledge. With the current liberalization of the criterion of empirical meaningfulness † the narrower positivism of the Vienna Circle has been definitively repudiated, and is being replaced by a ("hypercritical") realism. No longer do we identify the meaning of a statement with its method of verification. Nor do we consider the meaning of a concept as equivalent with the set of operations which in test situations enable us to determine its (more or less likely) applicability. Instead we distinguish the evidential (or confirmatory) basis from the factual content or reference of a knowledge claim. Early and crude forms of behaviorism identified mental states with their (*sic!*) observable symptoms. Embarrassment might then mean *nothing but* blushing. But refinements and corrections were introduced in due

* Cf. Carnap (62, 63); Skinner (320, 321); Wittgenstein (357).

† Cf. Carnap (64, 73); Hempel (149, 151); Feigl (105, 106, 109, 110, 112, 114, 116); Ayer (18); A. Pap (243, 246, 248). Also Grünbaum (139); Feyerabend (119).

course. Mental states were considered "logical" constructions based on observable behavior; and statements about mental states were considered logically translatable into statements about actual or possible behavior, or into statements (or sets of statements) about test conditions and ensuing test results concerning behavior. Mental traits were considered as correlation clusters of their (*sic!*) symptoms and manifestations, and so forth.

But even such a refined or "logical" behaviorism is now rejected as an inadequate reconstruction. It was realized that those behavioral test condition → test result conditionals are to be *derived* from the laws and postulates regarding central states. Such derivations or explanations have been eminently successful in the physical and in some of the biological sciences. In the atomic theory, or in the theory of genes, for example, it is becoming increasingly possible to derive the macro-regularities, regarding, e.g., chemical compounding, or Mendelian heredity from lawlike postulates and existential hypotheses. The central states of molar behavior theory (or the "factors" in the factor analysis of personality traits) are, however, unspecified as regards their neurophysiological basis. This is comparable to the early stages of the atomic theory when nothing was known about the mass and the structure of individual atoms, or to the early stages of the theory of heredity when Mendel's "units" were not as yet identified with the genes, located and spatially ordered in specific ways, within the chromosomes of the germ cells.

There is little doubt in my mind that psychoanalytic theory (or at least some of its components) has genuine explanatory power, even if any precise identification of repression, ego, superego, ego, id, etc. with neural processes and structures is still a very long way off. I am not in the least disputing the value of theories whose basic concepts are not in any way micro-specified. What I am arguing is that even *before* such specifications become possible, the meaning of scientific terms can be explicated by postulates and correspondence rules (cf. Carnap, 73), and that this meaning may later be greatly enriched, i.e. much more fully specified, by the addition of *further* postulates and correspondence rules.*

* For a defense of psychological theory without explicit reference to micro-levels, cf. Lindzey (200). The logic of theoretical concepts in psychology has been discussed in some detail by McCorquodale and Meehl (213); Feigl (113); Cronbach and Meehl (79); Ginsberg (133, 134); Maze (212); Seward (317); Rozeboom (283); Scriven (306).

After the recovery from radical behaviorism and operationism, we need no longer hesitate to distinguish between *evidence* and *reference*, i.e., between manifestations or symptoms on the one hand, and central states on the other; no matter whether or not central states are micro-specified (neurophysiologically identified).

The meaning of scientific statements consists indeed in their truth conditions. But "truth conditions" does not mean the same as "confirming evidence". (The only possible exceptions to this are the directly and completely confirmable singular statements regarding immediately observable situations.) A *theory* is required to tell us which observations form confirming evidence for scientific statements about matters inaccessible to direct observation. It is in the light of such theories that we can then specify how much support a given bit of evidence lends to a specified hypothesis.

In section V, I shall return to the crucial questions of reduction and identification. There I shall discuss the logical nature of the relation between mentalistically, behaviorally, and neurophysiologically characterized central states.

No elaborate arguments should here be required for a realistic interpretation of the statements about the "physical" objects of everyday life or of theoretical physics.* In the explanatory context (or the "nomological net") concepts pertaining to the unobservables are related to, but not identifiable with, the observables which constitute the evidential data for the confirmation of statements about the unobservables. For example, spectral lines, cloud chamber tracks, scintillations on screens, Geiger counter indications, etc. are the evidential data which, in a complete logical reconstruction, must be conceived as *nomologically* connected with the aspects of atomic and subatomic particles which they confirm. Less exciting, but logically analogous, is the analysis of statements of common life about ordinary (partly or wholly observable) objects. Here the perceived perspectives of mountains, trees, clouds, etc., or the instrument indications of air pressure, wind currents, air moisture, etc., are to be interpreted as evidence related to what is evidenced, by the geometrical-optical laws underlying the projections in visual perception, or the physical laws which explain the operation of barometers, anemometers, hygrometers, etc.

* Cf. B. Russell (288); R. B. Braithwaite (48); Kneale (179); L. W. Beck (24); Feigl (110, 111, 114).

6. The "meat" of an adequate solution of the mind-body problem will consist in a specific analysis of the characteristics and the relations between the attributes of the mental (especially the phenomenal) and the *physical* (specifically the neurophysiological). It should be clear from the outset that, if a complete solution of these problems is ever going to be achieved, it will arise out of a combination of the results both of scientific research and of philosophical analysis. In all these questions the two components are so intimately bound up with one another, that neglecting either of them seriously jeopardizes the whole endeavor. The philosophical aspects will be given a further analysis in the next section where I shall try to sort out the various meanings and the attached connotations of the terms "mental" and "physical". The most controversial, tangled and perplexing questions concern, of course, the distinctions made rightly or wrongly in the Cartesian and in the subsequent dualistic tradition between the *mental* and the *physical* in terms of the various alleged criteria listed in the accompanying table.

Mental	*Physical*
subjective (private)	objective (public)
nonspatial	spatial
qualitative	quantitative
purposive	mechanical
mnemic	non-mnemic
holistic	atomistic
emergent	compositional
intentional	"blind"; nonintentional

Practically all the perennial perplexities of the mind-body problem center around the listed contrasts. The dualists make *prima facie* an excellent showing. The more enlightened monists have always realized that any argument in favor of an identification (in *some* sense!) of the mental and the physical is faced with serious difficulties. Small wonder then that many of the more sophisticated analytic philosophers of the present age either embrace some form of dualism (usually parallelism), or else declare the issue between monism and dualism a pseudoproblem engendered by logical or terminological confusions. I do not share this outlook. In the following section I shall prepare the ground for an "identity" theory, and I shall present my formulation as well as my arguments in section V.

IV. Sorting Out the Various Meanings of "Mental" and "Physical". A Comparative and Critical Analysis

Much of the trouble with the mind-body problem arises out of the ambiguities and vaguenesses of the terms "mental" and "physical". Some of their connotations have been briefly indicated in the juxtapositions listed toward the end of the preceding section. I shall now attempt to analyze these and other meanings more closely, and to point out the merits and demerits of the various actual and possible usages of "mental" and "physical". Philosophers of the modern age clearly differ as to what constitutes the central core or (if there be such clarity!?) the *criteria* of the mental and the physical. Some philosophers fasten primarily upon one pair of distinctions, others on a different pair as of primary significance.

A. *"Subjective" versus "Objective"*. The juxtaposition of "subjective" and "objective" has been the source of endless and badly confused controversies throughout the ages. There is nevertheless something significant and worth preserving in this distinction. To say that a twinge of pain experienced by person A is "subjective" or "private" to him may simply mean that another person B, observing A's behavior, may *infer* A's pain, but does not *have* it, i.e. he does not directly experience it. Dentists do not have the toothaches of their patients. In one sense this is clearly analytic (tautological).* It is analytic for reasons analogous to those which make it self-contradictory to say that I am growing my wife's hair. (Schizophrenics are known to make assertions of this sort.) "I am eating with my wife's teeth" is merely funny, but not self-contradictory. "Dentists always suffer toothaches when their drill comes near the pulpa of their patient's tooth" is synthetic, but empirically false. "I am listening through my wife's ears" if meant literally (not metaphorically) is a border line case, depending on specific detailed interpretation. "I am enjoying Mozart's music exactly as my wife does" is synthetic and may even be rendered as "I have the *same* musical experience as does my wife." (Remarks about the two meanings of "same" will follow presently.)

The case is a trifle more complex for perception. Two persons sitting next to each other in the concert hall are said to hear the same music,

* This is now even admitted by Ayer (18) who had earlier (15) held it was synthetic. His earlier position was, however, incisively criticized by Pap (243, 248) and Wating (341).

or at a given moment the same tones or chords, produced by the pianist on the stage. But the facts of the case are really not fundamentally different from the first example. A does not *have* B's musical experience (or vice versa), even if their auditory discrimination, musical appreciation, etc., does not differ in any discernible way. They may be said to hear the same sounds, to be both equally impressed or thrilled by them; but common sense as well as scientific reasoning clearly indicates that their *experiences* are numerically different. Fundamentally this case does not differ from, e.g., the case of two thermometers immersed next to each other in the same glass of water. It is perfectly proper to say that these instruments indicate the *same* physical condition. It is also perfectly proper to say that the two thermometers not only indicate *but also* "have" the "same" temperature. (This is logically quite like saying that two marbles have the same color.) But it would be most improper and paradoxical to say that the *events* taking place in the one thermometer are *identical* with those in the other. This is not the place for a discussion of Plato's problem of the "one and the many." Suffice it to point out that the phrases "the same as" and "identical with" are ambiguously used. "Sameness" or "identity" may mean complete similarity, as in the case of the two musical experiences, or in the case of the two thermometric indications. But "sameness" or "identity" in other contexts means the numerical oneness of the individual referent of , e.g., two different names, or of two different unique characterizations (Russellian descriptions). I conclude then that it makes perfectly good sense to speak of the subjectivity or privacy of immediate experience. *Numerically* different but *qualitatively* identical (indistinguishable) experiences may be had by two or more persons, the experiential events being "private" to each of the distinct persons.

Terminological trouble, however, arises immediately when we take a scientific attitude toward direct experience and try to confirm, describe, or explain it "objectively." Is it not an "objective" fact of the world that Eisenhower experienced severe pain when he had his heart attack? Is it not a public item of the world's history that Churchill during a certain speech experienced intense sentiments of indignation and contempt for Hitler? Of course! What is meant here is simply that statements about facts of this sort are in principle *intersubjectively confirmable* and could thus be incorporated in a complete historical account of the events of our universe. To be sure, there are cases in which confirmation is *practically* outright impossible. The last thoughts and feelings of a man immediately before his death, especially in a case of complete paralysis, or of death occurring through electrocution, may be inferable only with scant reliability. But this is not different from the case (cf. Carnap, 67, p. 419f) of the confirmation of the electric charge of a specific raindrop that fell into the Pacific Ocean in a place far removed from any observers. Our current liberal formulation of the empiricist meaning criterion countenances all statements of this sort as perfectly meaningful. They do not fundamentally differ from other less difficult-to-confirm statements about, e.g., the "true thoughts" of a liar or play actor. Modern devices, such as the lie detector, and various clinical-psychological techniques enable us to test for such "private" events with increasing (though generally only relatively low) reliability.

The foregoing considerations suggest that the terms "subjective" and "private" at least in one of their commonly proper and serviceable usages are not to be considered as logically incompatible with "objective" or "public" in the sense of "in-principle-intersubjectively confirmable". Private states in this philosophically quite innocuous sense are then simply *central* states. (Whether these are ultimately to be conceived mentalistically or neurophysiologically may be disregarded for the moment; but this will of course be discussed quite fully later.) "Subjective" or "private" in this sense may then designate the referents of direct introspective reports, and it will be understood that these same referents may well be more indirectly characterized by descriptions involving inference from behavioral symptoms or test results of experiments on behavior. In those cases of subhuman animal behavior in which we don't hesitate to speak of experienced pains, gratifications, rage, expectations, etc., there are of course no introspective reports. But other aspects of such behavior are in many respects so similar to the human case that the ascription of raw feels is usually justified on the basis of analogy. Here again, the "private" means the central state which causally effects (or at least affects) the overt and publicly observable behavior.

The terms "subjective" and "objective" are indeed mutually exclusive if they are used in a quite familiar but different way. In designating some impressions, opinions, beliefs, value judgments, etc. as "subjective," we sometimes contrast them with the "objective truth," or "objective reality." If, e.g., my friend maintains that the room is cold,

I am inclined to argue with him by pointing to the thermometer (which reads, say, 74°); and perhaps by explaining his "impression" by the fact that he is too scantily dressed, or that he is sick, or suffers from anxieties, etc. Similarly in the more drastic cases of dreams, illusion, delusion, etc. we criticize some (interpretive) *judgments* as based on "*merely subjective*" evidence. And it should go without saying that disagreements in aesthetic value judgments may often be explained on the basis of individual or cultural differences. "De gustibus non est disputandum" is our final resort if no objectively justifiable standard can be agreed upon.* But wherever beliefs *can* be criticized as, e.g., "biased," "too optimistic," "too pessimistic," etc., there are standards, such as those of normal inductive inference, which may indeed justify the rejection or correction of such "all too subjective" convictions. Here "subjective" and "objective" are indeed incompatible, although of course there may well again be an "objective" explanation of the genesis of "subjective beliefs."

There is, however, also a philosophical and speculatively extended sense of "subjective" or "private". In this very special and highly problematic sense it is assumed that there may be subjective states which are in principle inaccessible to intersubjective confirmation. Here we had better speak explicitly of "*absolute* subjectivity" or "*absolute* privacy." It is *this* sense which is entertained in some of the more radically interactionistic forms of dualism. And it is this sense which by definition is incompatible with "objectivity" understood as intersubjective confirmability. As I have indicated before, I no longer insist that a doctrine involving the notion of *absolute* privacy is entirely devoid of *cognitive* meaning. But I am inclined to regard it as *scientifically* meaningless. To recapitulate: if the scientific enterprise is defined as necessarily requiring *intersubjective* confirmability of knowledge claims, then this follows immediately and quite trivially.

Now, I think it is an essential aspect of the basic working program and of the working hypotheses of science that there is nothing in existence which would in principle escape intersubjective confirmation. Allowances have already been made for the (sometimes) insuperable *practical* difficulties of even the most incomplete and indirect confirmations. But the optimistic outlook that inspires the advance of

* On the meaning and the limits of the justification of norms, cf. my essay (109).

science and informs its heuristic principles,* does not tolerate the (objectively) unknowable or "un-get-at-able." No matter how distant, complicated, or indirect the connection of scientific concepts with some (intersubjective) evidential bases may be, they would not be concepts of *empirical* science (as contrasted with the concepts of pure logic or mathematics) unless they could in some such fashion be "fixed" by "triangulation in logical space." The "fix" we are able to obtain may be as indefinite as it is when theoretical concepts (like those of the positron, the neutrino, or the meson in physics; that of the unit of heredity; or of memory traces; of the superego, of general paresis, or of schizophrenia in biology, psychology, or psychiatry) were *first* tentatively introduced by only very sketchily formulated postulates. The concepts of absolutely subjective or completely private data, however, are so conceived that they can be applied only on the basis of the direct experiences contained in a given stream of consciousness. A completely "captive mind" † might experience senselike qualities, thoughts, emotions, volitions, etc., but they would (*ex hypothesi*) not in any way, i.e., not even through weak statistical correlations, be connected with the publicly observable behavior or the neurophysiological processes of an organism.

While it is *difficult* to spin out this yarn in a consistent (let alone plausible) fashion, I do not think it impossible, in the sense that it would necessarily involve some self-contradictions. There are philosophers who have been concerned with an analysis of the meaning of the "continuance of a pure (immaterial) stream of experience after bodily death"; or with the problem of the "inverted spectrum" (Could pure sensory-like qualia like red and green, blue and yellow, be systematically interchanged for different persons, despite a complete similarity in their discriminatory and linguistic behavior, as well as in their neurophysiological processes?). Speculations of this sort were declared taboo and absolutely meaningless by the early logical positivists. They were compared with assertions about absolute space and time, the (Lorentzian) ether, the "bond" between cause and effect, or the existence of a metaphysical substance, over and above anything that could be verifiably known by science about spatio-temporal *relations*, coordinate

* Some philosophers rather speak of them as "metaphysical presuppositions"; for my criticisms of this interpretation of science cf. (110, 114).

† The idea and the phrase are Hilary Putnam's.

transformation, functional relations between observable properties or measurable magnitudes, or relations of compresence of various observable properties. There is no doubt that this positivistic cleansing of the Augean stables of metaphysics had a most salutary effect. But positivists (temperamentally often negativists), in their zeal and eagerness to purge the scientific enterprise of meaningless as well as superfluous elements, have often overshot their goal. A redressing of the balance has become necessary, and we pursue nowadays responsible analyses of, e.g., causal necessity * which are perfectly compatible with the basic antimetaphysical insights of Hume. Similarly as regards the notion of absolute privacy, it is illuminating to conceive it at least as a logical possibility, and then to state as clearly as feasible the reasons which can be adduced for rejecting the idea for our world as we have come to conceive of it in our science to date.

The notion of absolutely private data of experience, if such data are to be *described*, would require a purely phenomenal or absolutely private language. Such a language, by definition and *ex hypothesi*, could not serve as an instrument of communication. Even a completely solitary humanlike individual could not engage in audible (or visible, etc.) symbolic activities. Not even soliloquies in this physically expressible form would then be possible. For *ordinary* soliloquies, amounting to *more* than the unexpressed thoughts of a private thinker, are expressible, and the very expressions would provide (no matter how unreliable) clues to the "inner" thought processes.†

Now, of course, if by a "language" one means what is *customarily* meant by it (viz., an instrument or vehicle of intersubjective communication), then an absolutely private language is ruled out by definition. Language as we know it and use it is indeed not absolutely private in the sense explained. But *that* it is intersubjective reflects a basic empirical feature of our world, or at least a basic feature of our-world-as-we-conceive-it in common life and in science. But I must postpone discussion of the fuller implications of this feature until I present my dénouement of the "world knot" in the final section. For the present I submit that by a "language" one is not compelled to mean an instrument of interpersonal communication. The idea of the soliloquy (*intra-*

* Cf. Burks (59); W. Sellars (312, 313, 314).

† For an extremely lucid and succinct discussion of this point cf. P. E. Meehl (219).

personal communication) may be restricted and modified in such a manner that it refers to unexpressed and inexpressible thoughts. This preserves a sufficient "family resemblance" with the ordinary notion of language. Such an absolutely private language would still enable the solitary thinker silently to label the qualities of his direct experience and to think silent thoughts which have the logical form of declarative (singular, universal, etc.) statements. I could, for example, with the help of remembrance, think that extreme anger always gradually subsides, that a given tone-as-heard is increasing in intensity, etc. Knowledge thus formulated in a private language may well be called "*knowledge by acquaintance*." It is true that ordinary discourse entertains a much wider conception of knowledge by acquaintance. There it covers knowledge based on, and not essentially transcending, the observations (amplified by very moderate and limited inferences). Thus we can quite properly say that we know the properties of sticks and stones, of apples and oranges, the manners and mannerisms of our close friends "by acquaintance."

But this *ordinary* concept of acquaintance is not very sharply defined. Having actually seen Winston Churchill for a few seconds (when on July 10, 1954, he emerged from 10 Downing Street in London and entered his black limousine, holding his cigar and waving to the assembled small crowd), am I entitled to say that I know him "by acquaintance"? Would I know Churchill "by acquaintance" if I had seen him (or rather his image) only in the cinema newsreels? I leave it to the linguistically more sensitive and subtle Oxford analytic philosophers to decide these questions, or else to tell me that "knowledge by acquaintance" is a hazy notion, involving "slippery slopes" in various directions. (Anyway, the latter alternative is what I consider the best analysis of the *ordinary* usage of the term.)

For a *philosophical* usage of the term, however, I suggest that "knowledge by acquaintance" be understood as knowledge involving no inferential components—or, if this be chimerical, then knowledge involving only that minimum of inference which is present when only memory is utilized for the recognition of similarities and differences. It is in this sense that I could assert on the basis of acquaintance, "Ah, there is that peculiar smell again; I don't know what causes it, I don't even know how to label it; it is so different from any fragrances of flowers, perfumes, cigar smoke, burnt toast, tangerines, etc. that I can't even place it in a multidimensional scheme of the rank orders of smells; but I know I have experienced this smell before and I am (subjectively) sure I would recognize it in the future if I were to experience it again."

As I have said earlier, I make no claim for the infallibility of knowledge by acquaintance. Our world, being what it is, is such that corrections of subjective-experience judgments (knowledge claims made on the basis of direct acquaintance) are definitely possible from the vantage point of intersubjective observation. Moreover, it should require no reminder that I quite emphatically want to distinguish *acquaintance* from *knowledge by acquaintance*. "Acquaintance as such" (in the philosophically restricted sense) is to mean simply the direct experience itself, as lived through, enjoyed, or suffered; *knowledge* by acquaintance, however, is propositional. Knowledge claims of any sort may be valid or invalid; the statements which formulate such knowledge claims are either true or false. In the case of practically all knowledge claims which have scientific status, the confirmation of their truth is incomplete and indirect. Knowledge by acquaintance, however, is direct and complete in the following sense: it seems utterly inappropriate to ask someone what his evidence is for asserting that he, e.g., feels at the moment elated, depressed, anxious, dizzy, hot, cold, and so on through the various modalities and qualities.

The philosophically much misused and over-exploited term "self-evident" might well be redefined and restricted to just such reports of immediate introspection or self-observation. With this, possibly unwise, terminological suggestion I do not wish to imply any doctrine of "incorrigibility" in regard to such protocols of immediate experience. I grant that even such protocol statements may be in error; and not only for the generally admitted reasons such as possible slips of the tongue or the pen; but also because the predicates or relational words used in such statements, if they are what they are intended to be, viz. universals, presuppose for their correct application even in the "absolutely private" language (as fancied above) at least the reliability of memory. This alone would ensure that the same term is applied to an experienced quality of the same kind as before. Otherwise a protocol statement would simply amount to what would in effect be a first introduction of the predicate in question by stipulative-ostensive definition; * i.e., it

* The notion of "ostensive definition" is of course highly problematic. In contradistinction to what "definition" (explicit, contextual, recursive, abstractive, condi-

would amount to the resolution to use the same term on future occasions sufficiently similar to the present one. But on the occasion of the first use of a new term, the sentence containing it would be true only in the extremely restricted (very much like *analytic*) sense that "Λ," the label which I arbitrarily apply to the completely and incomparably new fragrance that I am just experiencing, designates the quality experienced during each of thc moments of its temporary occurrence of finite duration.

There are other uncertainties besides the ones mentioned in the use of (available) predicates for the qualia of immediate experience. Am I to describe the way I feel at a given moment as "happy", "joyous", "merry", "gay", "frolicsome", "blithe", "debonair", "light hearted", "buoyant", "bright", "animated", "gleeful", "hilarious", "jolly", or what?

It is time to draw some conclusions from this discussion. There is one meaning of "mental" in which it coincides with one meaning of "subjective". Let us call this meaning "phenomenal". In so calling it we may leave for later the question as to whether what is phenomenally given and phenomenally labeled is always also indirectly characterizable in an intersubjectively meaningful terminology. In any case we have isolated one contrasting (though not necessarily incompatible) pair of meanings for "mental" and "physical": the phenomenal (i.e., the subjectively confirmable) and the intersubjectively confirmable (i.e., the physical$_1$ in the terminology suggested above). The meaning of "mental" (synonymous with "phenomenal") looms large in introspective and phenomenological psychology. It is also prevalent in Gestalt-psychological descriptions of the configurations in phenomenal fields.

But in the "depth-psychological" statements of the psychoanalytic schools of thought, "mental" includes also subconscious, and some unconscious, states and processes. Since these are described largely with the help of metaphors and similes taken from the phenomenal (disregarding here those from the physical, e.g., mechanical, hydraulic, etc.) sphere, and inasmuch as detailed neurophysiological descriptions are

tional, coordinative, or even implicit) generally means, ostensive definitions cannot be rendered in speech, writing, or printing. "Definition" in its normal use always means specification of the meaning of some symbol by recourse to the meanings of other symbols. "Ostensive definitions" (if this phrase is to be retained at all) had therefore better be regarded as the establishment or acquisition of a linguistic habit, the inculcation of a bit of rule-governed linguistic behavior. In an absolutely private language it may amount to the stipulation of a rule which associates certain thoughts or images with specific other items or aspects of direct experience.

still lacking, it will be well to remember that the word "mental" as commonly employed by present-day psychologists covers both phenomenal and non-phenomenal states and events. The justification for the inclusion of the subconscious ("preconscious") and the unconscious in the realm of mind comes of course from some other attributes traditionally considered as criteria of mentality. We shall turn to those other attributes. The one which (for philosophical-historical reasons) will be taken up first is, however, not as essential in this connection as are some of the others further down the list.

B. Non-Spatial versus Spatial. The Cartesian distinction of *res cognitans* and *res extensa* still provides some philosophers of our age with what they consider one of their most powerful arguments in favor of a radical dualism. Mental states and events in contradistinction to physical bodies, so they claim, do not have a location, nor are they characterizable as having shapes or sizes. The apparent plausibility of this doctrine seems to me to derive mainly from (1) a confusion, and (2) inattention to phenomenal spatiality and its relations to physical spatiality. The confusion becomes evident in rhetorical questions asked by dualists, such as "where is the feeling of motherly love located?" "how many inches is it long?" "is it square or pentagonal?" I must confess I have little patience with these silly games. The feeling of motherly love is a universal, an abstract concept, and it makes as little sense to ask about its spatial location as it does in regard to the (physical) concept of temperature. We have here a category mistake of the crudest sort, a confusion between universals and individuals. It makes sense to ask about the location of individual things or events, but it is simply nonsense to ask about the location of a concept (properties or relations in abstracto).

The same sort of nonsense arises if, after hearing the sentence "the mental depression finally left him," someone asks, "Where did it go?" This sort of question can come only from taking the initial (metaphorical) statement as literally as we take "his wife finally left him." Concepts, whether they designate occurrent or dispositional properties, do not as such have spatial location; or rather it makes no sense to ascribe any such to them. But concepts which are constituents of singular (specific descriptive) statements * are applied to individuals. We

* I.e., sentences containing proper names or coordinates.

say "Anthony Eden felt depressed after the failure of the Egyptian campaign." In this case there is quite clearly a location for the feeling of depression. It is in the person concerned! The question of location becomes then more sensible, but logically also more delicate, if we ask it of individual mental states.

Using "mental" for the time being in the sense of "phenomenal", we had better—and without too much ado—introduce the indispensable distinction between phenomenal space(s) and physical space. I am perhaps not too acute in matters of phenomenological description but it does seem to me that my feelings and emotions pervade large parts of my body-as-I-experience it. William James has given us some striking illustrations of this. In the phenomenal field of the subject, specific feelings may be located at least vaguely or diffusely in some not very sharply delimited part of the organism. My feelings or sentiments of elation, depression, delight, disgust, enthusiasm, indignation, admiration, contempt, etc. seem to me to be spread roughly through the upper half or two-thirds of my body.

Sounds and smells at least in the usual situations of "veridical" perceptions seem to be partly outside, partly inside the phenomenal head. Colors are usually perceived as surface qualities of extradermal objects, or in the case of looking at the skin of one's own arms or legs, as surface qualities of those limbs. Colors seen when pressing one's eyelids (closed eyes) are vaguely located either immediately in front of one's eyes, or even inside them. Similarly musical sound images (especially in the eidetic's case) either appear inside one's head or seem to come from the outside as in a concert hall. The taste of an apple is clearly experienced within the mouth. The stars as seen on a cloudless night are tiny bright spots on a fairly distant dark background. These bright spots clearly have spatial relations to one another. A given small portion of the sky-as-perceived is an approximately plane surface with the twinkling stars distributed in certain constellations. If for the moment we may use the names of the stars as proper names for the bright spots in the visual field, we may well say that, e.g., Sirius is to the left and far below the three stars of Orion's belt. There is no question then that we are "acquainted" with the elements and relations in visual space.

A detailed discussion of the differing features of visual, tactual, kinesthetic, and auditory "spaces" is a task for phenomenal psychology. For our purposes it is sufficient to notice that "spatiality" means qualitatively quite different things for the various sense modalities. But *physical* space, in the sense in which the *science* of physics (including, of course, astronomy) understands it, is something radically different. The astronomers' measurements and inferential interpretations have provided us with an account of the three-dimensional array of the stars in "objective" space. This three-dimensional order is most properly considered as a conceptual system which can be only inadequately visualized or imaged phenomenally. I don't for a moment deny that in our rooms or in a landscape we *perceive* directly at least some of this three-dimensional order. (In the case of the stars, we don't.) But what is present in perception at any given moment is always a particular perspective and not the geometrical order which we must *assume* (together with certain laws of geometrical optics) in order to *explain* the peculiarities of any (or all) particular perspectives.

I shall not labor the obviously analogous case of time. Phenomenal time and physical time differ from, and are related to, each other very much like phenomenal space and physical space. Experienced durations may seem very long in the case of tiresome waiting, while time packed full with exciting events seems to "pass quickly." But the physically measured durations may be exactly the same. The *psychological* relativity of (phenomenal) time must of course not be confused with the (Einsteinian) *physical* relativity of simultaneity and duration which, in the nature of the case, is not directly observable at all.*

We conclude then that *mental* data have their own (phenomenal) kinds of spatiality; and that *physical* space is a theoretical construction introduced to explain the features and regularities of phenomenally spatial relations. The exact and detailed derivation, even only of the perspectival aspects of visual spatiality is a quite complex matter, involving geometrical, physical, psychophysical, and psychophysiological laws. Our arguments have so far disproved only the Cartesian contention that the mental is non-spatial. To put it very strongly, mental events as directly experienced and phenomenally described are spatial. Physical bodies geometrically characterized in their measurable positions, orientations, shapes, and sizes are not spatial (in the *visual*, or generally,

* Except, of course, for such cases as the traveling and returning twin brother, which, though strictly implied by the well-confirmed principles of Einstein's theory, has not been susceptible to direct check thus far (because of obvious practical difficulties).

phenomenal) sense at all. "Space" in the physical sense is an abstract theoretical ordering system. The reader who accepts my arguments may nevertheless maintain that the emphasized distinction between phenomenal and physical spatiality (and temporality) reaffirms all the more convincingly the dualism of the mental and the physical. My rebuttal of this contention will be given in the concluding sections. Suffice it here to suggest that if by "physical" we do not understand a kind, type, part, or aspect of reality, but rather a method, language, or conceptual system, then there is no room for a dualistic opposition of mental and physical events or processes, let alone substances.

C. *Quality versus Quantity*. Another time-honored distinction between the mental and the physical is made in terms of the qualitative and the quantitative. This distinction also is fraught with the danger of various confusions. A prima facie plausible argument maintains that, e.g., the qualities of colors-as-experienced, sounds-as-heard, odors-as-sensed, heat-intensities-as-felt, etc. are undeniably and fundamentally different from the quantitatively measurable wave lengths of light radiation, the frequencies and energies of sound waves, the chemical compositions of odorous substances, the mean kinetic energies of the molecules, etc. Of course, they are. But the argument misses the essential point. What the physicist measures are quantitative aspects of stimuli or stimulus patterns. These stimuli produce, under certain ("normal") circumstances, certain qualitatively characterizable sensations within the phenomenal fields. The familiar freshman's question, "Is there a sound when on a lonely island, with neither men nor beasts present, a tree falls to the ground?" is quickly clarified by the distinction between the sound waves (vibrations in the air) and sounds-as-heard. The dualistic argument would, however, be strictly to the point if it concerned the distinction between the sense-qualities-as-experienced and the "correlated" cortical processes in the brain of the experiencing subject. These cortical processes could be quantitatively described in a completed neurophysiology. Various more or less localized patterns of nerve currents ("firings" of neurons, etc.) would be the object of a "physical" description. Just which phenomenal qualities correspond to which cortical-process patterns has to be determined by empirical investigation. In our previous discussion of "conservanda" and "explicanda" we have not only admitted, but insisted upon, the *synthetic* character of the statements which formulate these correlations. Reserving fuller arguments for monism again for the final sections, a few preliminary critical observations are in order at this point:

(a) Purely phenomenal descriptions are generally not restricted to a merely qualitative form. Semiquantitative or rank-ordering ("topological") descriptions are possible at least among the qualities within each modality of experience. "My pain is increasing"; "this (sensed) blue is darker than that"; "my embarrassment was worse than any I had ever felt before"—these examples illustrate semiquantitative singular statements. Universal statements of this form can also be made, e.g., "Purple is more bluish than scarlet." "D is higher in pitch than C." Universal statements of this sort can be organized in topological arrays of one, two, three (or more) dimensions, as in the tone scale, the color pyramid, the prism of odors, etc. Moreover, there are cases of remarkable intersubjective agreement even in purely introspective judgments of the *metrical* relations of given qualities or intensities with each other. S. S. Stevens,* for example, found by careful experimentation that subjects agreed on what was the mid-point in a series of sounds of varying intensities. Shapes, sizes, distances, durations—all-as-directly-experienced are often susceptible to metrical estimates far surpassing in accuracy anything the uninformed might ever expect.

As regards the differences among such experiential modalities as colors, sounds, and smells, or between larger classes such as the sense qualities and the emotions, it must of course be recognized that they differ qualitatively from one another; and no merely quantitative distinction will serve as a criterion to characterize their different generic features. Dualists have tried to utilize this as an argument by asking, Why should there be more than one basic quality (or modality, for the matter of that), if all of the manifold phenomenal data are to be nothing but the subjective aspects of basically homogeneous brain processes? But the answer may well be that there are sufficient topographical, configurational, and quantitative differences even among those "homogeneous" neural processes.

(b) The magnitudes determined by physical measurement, and syntactically represented in scientific language by *functors*,† differ among themselves in a way that can hardly be called anything but "qualita-

* Cf. his article in the *Handbook of Experimental Psychology* (S. S. Stevens, ed.). New York: Wiley, 1951.

† Cf. Carnap (65, 68); Reichenbach (274).

tive". What else can we say about the differences between, e.g., mass, temperature, pressure, electric current intensity, electromotoric force, gravitational field intensity, etc.? What is it that is, respectively, indicated by thermometers, manometers, ammeters, voltmeters, etc.? I think it is entirely justifiable to speak of these scientific variables as *qualitatively* different. To be sure, they are not directly experienced qualities. But is there any good reason for restricting the term "quality" to the phenomenally given?

I conclude that the attempt to define "mental" and "physical" in terms of the distinction qualitative-quantitative begs the question. It makes perfectly good sense to speak of mental quantities and of physical qualities.

D. "Purposive" versus "Mechanical". Along with *direct experience*, it is perhaps *intelligence* which makes up the most important characteristic of the commonsense concept of mentality. And intelligence is usually and most basically characterized as the capacity of utilizing means toward the attainment of ends. One trouble with this characteristic is that common language is apt to describe as "intelligent" even the instinctive behavior of many animals. In the case of, e.g., social insects (termites, ants, bees, etc.) the behavior is stunningly purposive, highly organized, and intricate; and yet we hesitate to ascribe sentience or subjective experience (raw feels) even only remotely resembling our own to these entirely different organisms. Moreover, the current *scientific* use of the word "intelligence" tends to be restricted to those evolutionary levels and species in which learning combined with ingenious (inventive) and symbolic behavior plays a dominant role. Pigeons, rats, cats, dogs—those favorite laboratory animals of the behavioristic psychologists—show (in each species) marked individual differences in the speed and the scope of their learning. Anthropoid apes, like the chimpanzees, are famous (ever since W. Köhler's original experiments) for their inventiveness—in addition to their commonly known capacities for imitation. Genuinely *linguistic* behavior, involving syntactical, semantical, and pragmatic features, seems to be restricted to *homo sapiens*; the so-called language of the bees (which is apparently instinctive and lacking in syntactical and semantical flexibility) does not seem to be an exception.

If intelligence or just purposiveness were chosen as the sole criterion of mentality, then it would be hard to draw a sharp line anywhere within the realm of organic life. Even in the kingdom of plants we find processes whose teleological characteristics are not fundamentally different from the features of purposive behavior in the lower animals. Of course, if one deliberately makes the (often suggested and no doubt helpful) distinction between two types of teleology, one of them involving *conscious* aims, and the other excluding them, and designates only the former as "purposive," then the empirical evidence suggests (but does not force upon us) the decision to call "intelligent" only the behavior of the higher animals, or perhaps to restrict the label "intelligence" to human beings (i.e., if and when they behave in a genuinely *sapient* manner).

It becomes clear then that the scope of the two criteria (*sentient* and *sapient*) is not necessarily the same. The two concepts are not coextensive. The situation has been further complicated in our age by the construction of "intelligent" machines. Logical reasoning, mathematical proofs and computations, forecasting, game playing, etc. are all being performed by various and usually highly complex electronic devices. Here the temptation to ascribe "raw feels" becomes even weaker than in the case of the lower animals.* Inductively it is plausible that sentience requires complex organic processes.

Descartes was perhaps not completely wrong in restricting mentality to human beings. If "mind" is understood as the capacity for reflective thought, then indeed we may have reason to deny minds (in *this* sense!) to animals (and perhaps even to electronic computers!). The issue is difficult to decide, because the connotations of "reflective thought" are numerous and indefinite. But if it connotes a *conjunction* of sentience, learning capacity, spontaneity (free choice), purposiveness (in the sense of goal directedness), original inventiveness, intentionality (in the sense of symbolic reference), and the ability to *formulate rules* of behavior (practical, moral, linguistic, etc.), then *mind* (in this sense) is clearly the prerogative of man.

All the foregoing considerations need not disturb us. They merely lead to the scarcely surprising conclusion that the term "mental" in ordinary and even scientific usage represents a whole family of concepts; and that special distinctions like "mental$_1$", "mental$_2$", "mental$_3$",

* Cf. however, the remarkable and stimulating discussion of the robot problem by Scriven (304). We shall return to this issue in connection with the scrutiny of the analogy argument for "other minds" in section V.

etc. are needed in order to prevent confusions. (We shall return to a brief discussion of "intentionality" in subsection *F*.)

As far as the original distinction of *purposive* versus *mechanical* is concerned, it scarcely helps in the definition of the mental versus physical distinction. If "purposive", despite our warnings, is taken as synonymous with "teleological", then we have a distinction, which, though it becomes rather irrelevant to the mental-physical issue, is not useless in the natural sciences and in technology. But then it can no longer be considered as either sharply exclusive, or as particularly enlightening. The flow of a river toward the sea is a mechanical and non-teleological phenomenon, but the functioning of servomechanisms is mechanical as well as teleological, and the functioning of the heart is teleological and presumably "mechanical" in the same (wider) sense in which complex servomechanisms operating by negative feedback are regulative *physical* devices. In short, the phrase "teleological mechanisms" in our age of cybernetics is no longer a contradiction in terms.

E. "Mnemic", "Holistic", "Emergent" versus "Non-Mnemic", "Atomistic", "Compositional". This bundle of contrasts has often been associated with the distinction of the mental and the physical. Fortunately, except for one facet of the emergence issue, discussion can be quite brief. The *mnemic* as a criterion of mind was stressed especially by Bertrand Russell. But long before him, the physiologist Ewald Hering (and his disciple Semon) considered the mnemic as a general property of all organic matter. Even in inorganic matter there are more or less permanent modifications of dispositional properties which can be effected by various influences. Certain features of elasticity and of magnetic hysteresis are "mnemic" in this sense. And of course the storage of information in present-day computing machines clearly shows that mnemic features, just as the "purposive-intelligent" features, need not coincide with mentality in the sense of sentience or awareness.

The holistic aspects of the phenomenal fields were brought to the fore by the Gestalt psychologists. But almost from the beginning, this school of thought (especially ever since W. Köhler's book on *Physical Gestalten*, 1920) emphasized the idea of the isomorphism of phenomenal with neurophysiological configurations. Thus again, without the addition of the criterion of immediate experience we do not obtain a distinction between the mental and the physical configurations or "organic wholes" or "dynamic Gestalten."

Inseparably connected with holism and the Gestalt philosophy is the doctrine of *emergence*. The old slogan "the whole is greater than the sum of its parts" has of course no very clear meaning. Much of its obscurity is due to the lack of a definition of the phrase "the sum of the parts". Recent analyses * of the still controversial significance of "organic wholeness" and of "emergent novelty" have contributed a great deal to the clarification of the issues. There is no imperative need for us to enter into details here. It will be sufficient for our concerns to realize that in modern natural science no sharp distinction can be made between *resultants* (as in the composition, i.e. vectorial addition of forces or velocities) and *emergents*. In the explanation of the properties and the behavior of complexes and wholes we always need laws of composition—be they as simple as the straightforward arithmetical addition of volumes, masses, electric charges, etc., or slightly more complicated as is vector addition, (or just a trifle more involved as is the relativistic "addition" formula for velocities), or extremely complex as are the so far not fully formulated composition laws which would be required for the prediction of the behavior of organisms on the basis of a complete knowledge of their microstructure and the dynamic laws interrelating their component micro-constituents.

Modern quantum physics, on a very basic level, employs laws which have "organismic" character, as for instance the exclusion principle of W. Pauli † which holds even for single atoms. It is conceivable that much of what is called "emergent novelty" on the chemical and biological levels of complexity may ultimately be explained in terms of the organismic or holistic features of the laws of atomic and molecular dynamics; and that, given those basic micro-laws, the only composition laws (which scientists often take for granted like "silent partners") are simply the postulates and theorems of geometry and kinematics. This is indeed my own, admittedly risky and speculative, guess; that is to say, I believe that once quantum dynamics is able to explain the facts and regularities of organic chemistry (i.e. of non-living, but complex compounds) it will in principle also be capable of explaining the facts and regularities of organic life. But no matter whether these conjectures

* Schlick (299); Nagel (232, 235); Henle (153); Bergmann (28, 34); Hempel and Oppenheim (152); Rescher and Oppenheim (277); Pap (244).

† Cf. the clarifying discussion by Margenau (208); and the stimulating, but perhaps somewhat speculative, ideas of Kaila (169).

prove correct or incorrect, emergent novelty from a logical or methodological point of view simply means the impossibility of the derivation of the laws of complexes ("wholes") from the laws that are sufficient to predict and explain the behavior of their constituents in relative isolation. Thus, the laws that are sufficient to account for the motion of free electrons (as in cathode rays, and traversing electric or magnetic fields) are clearly insufficient to account for the behavior of electrons when they are constituents of atoms.

It stands to reason, that in order to "glean" (i.e., to ascertain) the laws of nature, scientists can't afford to stop their investigations on a very low level of complexity. In some cases we are lucky in that from such a very low level of complexity upwards to higher complexities of any degree, no new *physical* laws (but only geometrical composition laws) are required. This holds, for example, for the law of the lever which remains applicable even for the most complex system of pulleys. It also holds for the law of gravitation and the laws of motion (both in their Newtonian form). The "many bodies problem" is unsolved only in the mathematical sense that no single set of simultaneous equations has as yet been found for the prediction of the motions in complex star systems. But successive approximations can be computed to any desired degree of accuracy. In other cases (as with the behavior of electrons) we could never glean all the relevant laws below a certain level of complexity. And I have admitted (in section II) that it is always *logically* conceivable that our scientific theories may have to be amended and enriched by the introduction of new basic concepts (variables), and this is of course tantamount to the introduction of new (lawlike) postulates and/or existential hypotheses.

We have seen that the mnemic, teleological, holistic, and emergent features are not adequate as criteria of mentality, because these features characterize even inorganic structures and processes. Emergence as conceived by most dualists, however, refers to the evolutionary novelty and the (physical$_2$) underivability of *sentience* or *raw feels*. The whole issue therefore turns again upon the criterion of *subjective experience*. The issue can be brought out by questions such as the following: Suppose we could predict the detailed chemical structure of an entirely new perfume which will be manufactured in Paris in the year 1995. Suppose, furthermore, that we could equally exactly predict the neurophysiological effects of this perfume on the mucous membranes of a human nose, as well as the resulting cortical processes in the person thus smelling the perfume. Could we then also predict the quality of the experienced fragrance? The usual answer to this question is in the negative, because it is assumed that the fragrance in question will be an "emergent novelty." But behaviorists, and physicalists generally, need not take such a pessimistic view. For given the presuppositions of our questions it should also be possible to predict the answers to questionnaire items like "Is the fragrance more similar to Chanel 5 or to Nuit d'Amour?" That is to say, we should be able to predict the location of the quality in the topological space of odors, provided we have a sufficiency of psychophysiological correlation laws to make this particular case one of interpolation or (limited) extrapolation.

The issue can however be made more poignant if we are concerned with the prediction of qualities within an entirely new modality. In the case of the congenitally blind who by a cataract operation suddenly attain eyesight, the experience of colors and (visual) shapes is a complete novelty. Suppose that all of mankind had been completely blind up to a certain point in history, and then acquired vision. Presupposing $physical_2$ determinism we should (according to my basic conjecture) in principle be able to predict the relevant neural and behavioral processes, and thus to foretell all the discriminatory and linguistic behavior which depends upon the new cortical processes (which correspond to the emergent, novel qualities of experience). What is it then that we would not or could not know at the time of the original prediction? I think the answer is obvious. We would not and could not know (then) the color experiences *by acquaintance*; i.e., (1) we would not *have* them; (2) we could not *imagine* them; (3) we could not *recognize* (or *label*) them as "red", "green", etc., even if by some miracle we suddenly *had* them, except by completely new stipulations of designation rules.*

I conclude that the central puzzle of the mind-body problem is the logical nature of the correlation laws connecting raw feel qualities with neurophysiological processes. But before we tackle this difficult question, a glance at one more issue is required.

F. "Intentional" versus "Non-intentional". The mental life of (at least) the adult *homo sapiens* is characterized by the capacity for awareness—in addition to the occurrence of mere raw feels. (We credit some

* Cf. Pap's discussion of absolute emergence (244).

animals and certainly young children with the latter in any case.) To *have* an experience, and to *be aware* of having it, is a distinction which I think cannot be avoided, even if in a given case it may be very difficult to decide whether awareness actually supervened. This is one of the notoriously difficult questions of phenomenological description. But assuming the distinction, it is fairly plausible that awareness is impossible without some sort of symbolism, even if it be the "silent" symbolism of imagery or (if there be such) of imageless thought. It is here where the idea of "*intention*" (not in the sense of purpose, end-in-view, or resolution, but) in the sense of *reference* becomes essential.

I shall try to show that the scientifically relevant issues regarding interactionism versus parallelism (or epiphenomenalism) should be carefully separated from the philosophical issues which stem from the "intentional" features of mind, stressed by Brentano and the phenomenological schools of thought. According to this point of view the most fundamental difference between the mental and the physical consists in the fact that the mental life consists of *acts directed upon objects*, no matter whether these objects exist in the world, or are pure concepts, or figments of the imaginations. It is true that dualism in the Cartesian tradition has emphasized the intentional as well as the raw feel features of mind. The mind-body problems in the larger sense therefore have customarily included such questions as, Can we give a physical (1 or 2, in this case) account of how thoughts, beliefs, desires, sentiments, etc. can be *about* something? Can we give a naturalistic translation of the language of reasoning as it occurs in arguments, i.e., discourse in which we give *reasons* intended to support knowledge claims, or value judgments? I think it has become increasingly clear * that the answer must be in the negative; but not because human behavior involving "higher thought processes" is not in principle capable of physical (at least physical$_1$) explanation and prediction; but rather because the problem is one of the logical reducibility or irreducibility of discourse involving *aboutness* (i.e., intentional terms), to the language of behavioral or neurophysiological description. Now it seems fairly obvious that such discourse, just like discourse involving *oughtness* (i.e., normative discourse) is *not* logically translatable into purely factual statements. The relation of designation (formalized in pure

* Cf. especially Wilfrid Sellars (310, 311).

semantics) is not an empirical relation, but a construct of semantical discourse.

Personally, I therefore consider the problem of intentionality not as part of the psycho-physical but rather as a part of the psycho-logical problem, i.e., as part of the relation of psychological to the logical forms of discourse. This becomes even more evident because, assuming the ultimate possibility of a full neurophysiological account of behavior (including linguistic behavior), we should then have the problem of relating the *physiological* to the logical forms of discourse. If many writers permit themselves nowadays to speak of "thinking machines" (electronic computers, chess playing machines, etc.), then it is equally justified to pursue the problem of the relation between the *mechanical* (or the electrical) and the logical. In the case of the machines, it is ourselves who have built them in such a way that in their functioning they conform to certain rules of logical, mathematical, or semantical operations. In the case of human beings we have nervous systems which through education and training acquire the dispositions toward certain types of symbolic behavior which in actual operation then is more or less in conformity with certain rules.

But the abstract statement of a rule is not to be confused with the formulation of the (statistical) empirical regularity of the symbolic behavior. An illicit inference or a computation mistake is a *violation* of a *rule*, it is not an instance which would disconfirm a law of behavior. The recent phase of the clarification of these issues was in essence initiated by Husserl and Frege in their critique of *psychologism*, i.e., of the confusion of logical with psychological discourse. The pan-empiricist position of, e.g., John Stuart Mill who regarded logical truths as on a par with the truths of the natural sciences, was thus effectively and definitively refuted. Later, very much needed refinements of the anti-psychologistic position were added by Carnap (65, 68, 69, 71, 72), and a full study of the logical status of rules and rule-governed behavior has been contributed by W. Sellars (*loc. cit.*).

No matter what the most clarifying analysis of rule-governed symbolic behavior in its relation to the rules as such may turn out to be, there can be no doubt that if physical (at least $physical_1$) determinism is to be maintained, the following will have to hold: A person's brain state when thinking, e.g., about Napoleon's defeat at Waterloo must qualitatively or structurally differ from the brain state of the same person (or, for that matter, of other persons) when thinking about Caesar's crossing of the Rubicon. *This* aspect of the psychology and physiology of thought is definitely relevant for our problem.

V. Mind-Body Identity. Explications and Supporting Arguments

With due trepidation I shall now proceed to draw the conclusions from the preceding discussions, and to present the dénouement of the philosophical tangles. There are many points on which I have sincere and serious doubts. There is yet a great deal of analytic work to be done on several puzzling aspects for which I can at present only sketch the sort of solution which seems to me especially plausible.

A. *Review of the More Basic Meanings and Connotations of "Mental" and "Physical". Conclusions regarding their Respective Merits and Demerits.* The surveys and discussions of the preceding sections have paved the way for a summary and systematic appraisal of various characteristics which have been proposed as defining criteria of the mental and the physical. Outstanding candidates among the criteria of mind are (1) direct experience and (2) intelligence. "Direct experience" is synonymous with one sense of "subjectivity", viz. sentience, raw feels, or phenomenal givenness. "Intelligence" connotes learning capacity, purposive (goal directed) behavior and—on the human level—intentionality (symbolic behavior). Although the two criteria have in fact a certain area of coincidence, this coincidence (or overlap) is not a matter of logical necessity. By and large then, the two criteria of mentality define two entirely different concepts.

"Mind" as we have come to suspect all along, is an ambiguous term, or at best a group of concepts with family resemblances (in Wittgenstein's sense). The major components of the connotation of "intelligence" may be attributed not only to the higher animals but also to the "thinking machines" which we generally consider not only as lifeless but also as devoid of sentience. Direct experience, on the other hand, may well be attributed to some of the lower animals, babies, idiots, and to the severely insane; but in each of these classes at least some, if not all, of the marks of intelligence are lacking. Furthermore, it is customary in contemporary psychology to classify the unconscious (deeply repressed) traumata, anxieties, wishes, conflicts, etc. as *mental*. This again indicates that direct experience is not the criterion here, even if—according to the psychoanalytic doctrine—deeply repressed matters are *potentially* conscious, in that they can be brought to the fore of awareness by special techniques. Hypnotic and posthypnotic phenomena also often involve deeply unconscious processes, which because of their other similarities with the conscious processes are unhesitatingly classified as mental.

One might suppose that the term "physical" (to which we have paid thus far only sporadic attention) is much more definite in meaning than the term "mental". Unfortunately, the contrary is the case. There are some superficial and entirely inadequate definitions of "physical" which need only be mentioned in order to be promptly dismissed. For example, to define "physical" as the "outer" aspect (in contradistinction to the "inner" mental life) is to use misleading metaphors. "Inside" and "outside", "internal" and "external" have a good clear meaning in ordinary usage. What is literally inside, e.g., the skin of a person is most of his body (i.e., the body minus the skin) and that's "physical" in at least one very good sense of the term. After all, anatomy and physiology are concerned with the physical structure and the functions of organisms. *Inside* the skull is the brain of man, and that is "physical" in the same well understood sense.

Similarly unhelpful is the definition of the "physical" as the *mechanical*-compositional, as contrasted with the *purposive*-holistic. We have already repudiated this sort of definition-by-contrast, by pointing out that "mechanical" in the strict sense of "characterizable by the concepts and laws of Newtonian mechanics" designates only a narrow subclass of the class of *physical* events or processes, using "physical" (comprising also electrodynamic, relativistic, and quantum-theoretical characteristics) in the sense of modern physics. And if by "purposive" we mean no more than by "teleological" and "holistic", then there are innumerable teleological mechanisms, many of them with typical features of organic wholeness, both in nature and among the artifacts of technology. If "purposive" is understood in the narrower and more fruitful sense, then it involves intelligence (and this, on the human level, includes intentionality).

But the fact *that* there are (human) organisms functioning intelligently and displaying (symbolic) behavior which indicates intentional acts is describable in an intersubjective ("physical_1") manner and therefore again does not support a definition-by-contrast between the physical (in this case physical_1) and the mental. It remains true, however, that among the objects and processes describable in $physical_1$ terms, there are differences at least of degree (often of very considerable degree) if not of a fundamental, evolutionary-emergent type, as between the structure and the dynamics of electrons, atoms, molecules, genes, viruses, and unicellular and multicellular organisms. The tremendous differences between, e.g., a simple inorganic structure and a human being are therefore not in the least denied. As Castell (74) puts it, the solar system and an astronomer thinking about it, are in many essential respects very dissimilar indeed. (But the dualistic conclusions drawn by Castell seem to me nevertheless *non sequiturs.*)

The foregoing considerations suggest some of the more fruitful definitions of "physical". "$Physical_1$" may be defined as the sort of objects or processes which can be described (and possibly explained or predicted) in the concepts of a language with an intersubjective observation basis. This language or conceptual system is—in *our* sort of world—characterized by its spatio-temporal-causal structure. This is so fundamental a feature of our world that it is extremely difficult to imagine an alternative kind of world in which intersubjectivity is *not* connected with this feature. One can understand, but need not concede, Kant's contentions regarding the synthetic a priori character of this "presupposition." * The concept of "$physical_1$" is closely related to but by no means equivalent with one of the primary meanings of "physical" in ordinary language, viz. observable by sense perception. In its most natural usage "observable by sense perception" clearly comprises the solid and liquid objects of our environment; it includes of course our own bodies; it includes a trifle less clearly the air (which can be felt if it moves with sufficient speed; or other gases if they can be smelled); it includes less obviously some of the dispositional properties of various sorts of matter (such as their hardness, elasticity, solubility, fusibility, etc.); and it scarcely includes electric or magnetic fields, atoms and electrons, or the secret thoughts of other persons.

But in *one* usage "observable by sense perception" does comprise the feelings, emotions, and even some of the (dispositional) personality traits of other persons. For example, we say, "I could see how disappointed he was." "I can see that he is a depressive person," etc. But these are usages which from the point of view of logical analysis are perhaps

* For a critique of this rationalistic position, cf. Pap (242); Nagel (233); Reichenbach (275); Feigl (114).

not fundamentally different from the case of a physicist, who (looking at a cloud chamber photograph of condensation tracks) says, "Here I see the collision of an electron with a photon." Such (extended) "observation statements" urgently demand a logical analysis into their *directly* verifiable, as contrasted with interpretive and inferential components. Logical analysis, pursuing as it should, an epistemological reconstruction, must therefore be distinguished from phenomenological description.

From the point of view of a phenomenological description, the "preanalytic data" of the clinical psychologist contain his direct impression of (some of) the personality traits of his clients; just as the experienced physician's judgments may be based on his direct impression of the disease (diabetes, multiple sclerosis, Parkinson's disease, etc.) of his patient. Phenomenological description is a subtle and interesting matter, but philosophically much less relevant than it is often supposed to be. By a little exercise of our analytic abilities we can, and for epistemological purposes we must, separate the directly verifiable situation (the patient is very slow in all his movements, hangs his head, speaks with a very low voice; or: he has dry skin; his breath has a fruity smell; his hands tremble; etc.) from the inferential interpretations, i.e., the conclusions regarding his mental or physical illness.

Inasmuch as the use of terms like "psychoneurosis" is established, and diagnoses of psychoneuroses can hence be confirmed, on an intersubjective basis, the *concept* of psychoneurosis is evidently a physical_1 concept. At least partial *explanations* of the behavior and the subjective experience of psychoneurotics have also been given on a physical_1 (roughly: behavioristic) basis. We can plausibly explain neurotic dispositions by tracing them causally to the childhood situations of the patient (not necessarily neglecting some of his biologically inherited constitutional traits). And we can predict his anxieties, depressed moods, etc. on the basis of such intersubjectively confirmable information as, e.g., about a preceding period of highly "id-indulgent," overbearing, or hostile behavior. These "physical_1" explanations do not differ fundamentally from explanations of, e.g., the growth of plants or the behavior of lower animals. That a plant grows poorly may be explained by the sandy soil in which it is rooted, the lack of rainfall, etc. The behavior (or some aspects of it) of an amoeba may be explained by the thermal and chemical conditions of its immediate environment.

The distinction between *psycho*neuroses and "physical" nervous disorders originates from the same commonsense considerations that have traditionally led to the contrast of "states of mind" and states of the body. No matter whether normal or abnormal processes are concerned, whenever scientifically or philosophically innocent people speak of something as being "in the mind" or "merely in the mind," this means apparently that it is not directly accessible to sensory observation. But, it is also positively characterized by the fact that these "states of mind" can (usually) be reported by those who *have* them, and that they can (sometimes) be influenced by talking. Sticks and stones cannot be made to move by merely talking to them.* Persons (having minds!) can be made to do things by suggestions, propaganda, requests, commands, etc., often by just giving them certain bits of information.

But important and interesting as is this sort of difference, in its scientific aspects it no longer establishes a fundamental difference between *inanimate things* and *minded persons.* Modern robots have been constructed which emit information about their "inner" (physical!) states, and they *can* be made to do things by speaking to them. But if intellectually acute and learned men † discuss seriously the problem as to whether robots really have a mental life (involving thoughts and/or feelings), there must be a question here that clearly transcends the obviously *scientific* and *technological* issue as to whether robots can be constructed which in their behavior duplicate all essential features (of course, one must ask: which ones and how completely?) of human behavior. If by "thinking" one means a performance starting with "input" premises and culminating in "output" conclusions which are deductively or inductively implied by the premises, and if the performance consists (at least) in certain observable relations between *input* and *output*; then there is no doubt that certain types of robots or computers do think. If one means by "feeling" what the logical (or illogical?) behaviorists mean, then it is at least conceivable (cf. Scriven, 304) that there might be machinelike structures (artificially made, or even naturally existing on some other stars) which behave (respond, etc.) in every way *as if* they had feelings and emotions.‡ Here the question is

* This still seems safe to assert even in view of the alleged but highly questionable "facts" of psychokinesis.

† Cf. Turing (338); MacKay (216); Spilsbury (326); Scriven (304).

‡ The question in this form is by no means new. William James discussed it in his *Principles of Psychology* (Vol. I) by means of the example of the "automatic

clearly of the same logical nature as the queries: "Do butterflies feel?" "Do fishworms, when put on the hook, feel pain?" "Do plants have feelings?" "Do human embryos, four months old, have any direct experience?" I shall try to clarify the nature of these questions in the following subsection. For the moment it must suffice to point out that here we have to do with the distinction between "mental" (in the sense of *sentience*) and *physical*$_1$. *Intelligence,* in contrast to sentience, is clearly definable in physical$_1$ terms. But as to whether *sentience* is so definable is perhaps the central perplexity among the mind-body puzzles.

But now to complete our analysis of the meanings of "physical": We have distinguished "physical$_1$" and "physical$_2$". By "physical$_1$ terms" I mean *all* (empirical) terms whose specification of meaning essentially involves logical (necessary or, more usually, probabilistic) connections with the intersubjective observation language, as well as the terms of this observation language itself. Theoretical concepts in physics, biology, psychology, and the social sciences hence are all—at least—physical$_1$ concepts. By "physical$_2$" I mean the kind of theoretical concepts (and statements) which are sufficient for the *explanation,* i.e., the deductive or probabilistic derivation, of the observation statements regarding the inorganic (lifeless) domain of nature. If my conjecture (discussed above) is correct, then the scopes of *theoretical* "physical$_1$" and "physical$_2$" terms are the same. *If,* however, there is genuine emergence, i.e., logical underivability, in the domains of organic, mental, and/or social phenomena, then the scope of "physical$_2$" terms is clearly narrower than that of "physical$_1$" theoretical terms.

Within the category of "physical$_1$" terms, it is clearly important to distinguish observation terms from theoretical terms; and among the latter several levels may methodologically, if not logically, be distinguished. For example, the concepts of classical thermodynamics form one level, and the concepts of statistical or molecular mechanics (in terms of which those of thermodynamics, with certain modifications,

sweetheart." He was severely criticized by E. A. Singer (319) who, ironically enough, appealed to James' own principle of pragmatism (derived from Peirce's meaning criterion which anticipated the essentially equivalent later operationist and logical-positivist formulations of the criterion). But E. A. Singer in turn was incisively criticized by D. S. Miller (224), who many years later (226) attacked on the same grounds the much more subtle linguistic behaviorism expounded in Gilbert Ryle's *The Concept of Mind.*

can be defined) form a "higher" level. The concepts of molar behavior theory are related analogously to those of the higher level of neurophysiology; and so on *mutatis mutandis,* throughout the various fields of scientific theories.

We conclude that to say "x is physical" is highly ambiguous. There is first the obvious distinction between the physical *languages* (physical language designators) and physical *objects* (physical language designata). This distinction carries through the two further distinctions and does not, for our purposes, require elaborate discussion. To illustrate, an electromagnetic field, just as the planet Jupiter, are *designata* of physical language *terms.* However, the *observation* terms of the physical$_1$ language serve also as the evidential basis of the physical$_1$ or physical$_2$ theoretical languages. *Theoretical* terms are here conceived as not *explicitly* definable on the basis of observation terms (cf. Carnap, 73; Feigl, 110; Sellars, 315), but as specified by postulates and by correspondence rules relating them to the terms of the observation language. And, to restate this in different words, *if* there is no genuine emergence in the logical sense above the level of lifeless phenomena, then there is no basic distinction between the theoretical terms of the physical$_1$ and physical$_2$ languages. That is to say that the theoretical terms of biology and psychology are explicitly definable on the basis of the theoretical concepts of physics in the same sense as the theoretical terms of chemistry (e.g., the chemical bond) are nowadays explicitly definable on the basis of the theoretical terms of the physical$_2$ language (i.e., of the atomic and quantum theories).

The central questions of the mind-body problem then come down to this: are the concepts of introspective psychology—relating to phenomenal data or phenomenal fields—definable on the basis of physical$_1$ theoretical terms, and if so, are they also definable on the basis of physical$_2$ (theoretical) terms? The first question is a matter for *philosophical* analysis. The second question is, at the present level of scientific research, undecided, though my personal (admittedly bold and risky) guess is that future scientific progress will decide it affirmatively. We turn now to a discussion of the first question primarily, but occasional remarks about the second question will also be ventured.

B. The Inference to Other Minds. Behaviorism and phenomenalism display interesting similarities as well as fundamental differences. According to logical behaviorism, the concepts of *mental* states, dispositions, and events are logical constructions based on (physical$_1$) characterizations of behavior. According to the more recent formulations of physicalism (Feigl, 113, 116; Carnap, 73; Sellars, 315) the "logical construction" thesis is inadequate and has to be replaced by an analysis in terms of postulates and correspondence rules. Very simply and very roughly, this means—in the material mode of speech—that for physicalism mental states are inferential ("illata," cf. Reichenbach, 273). Contrariwise, modern phenomenalism (Carnap, 60; Ayer, 12; Goodman, 135) had maintained that the concepts of *physical* things, states, dispositions, and events are logical constructions based on concepts designating the phenomena of immediate experience. And in the "revised" version of phenomenalism, i.e., a genuinely *realistic* epistemology based on phenomenal data, a doctrine which should not be called "phenomenalism" at all, the concepts of physical objects are inferential ("hypothetical constructs," "illata"). But this doctrine is in many of its tenets consonant with classical critical realism (von Hartmann, Külpe, Schlick, R. W. Sellars, D. Drake, C. A. Strong, J. B. Pratt, A. O. Lovejoy, G. Santayana). In contradistinction to critical realism, there is the earlier doctrine of *neutral monism* developed by the neorealists, especially E. B. Holt and Bertrand Russell (before his later critical realism), and historically rooted in the positivism and empiriocriticism of Hume, Mill, Mach, and Avenarius. Russell (284, 287) was the primary influence in Carnap's early epistemology (60, 61); and this sort of neutral monism was also adopted in prefatory philosophical remarks of some psychologists like E. C. Tolman (336), C. C. Pratt (260), and others.

The distinctive mark of neutral monism is a conception of the "given" which (1) is subjectless, i.e., it does not allow for the use of the personal pronoun "I"; and (2) is "neutral" in the sense that the given is characterizable as *neither* "mental" *nor* "physical." It maintains that both mentalistic concepts (the concepts of psychology) and physical concepts (those of physics) are logically constituted out of the more basic concepts designating neutral data. Psychology and physics are here understood as more or less systematic knowledge both on the level of common life, and on the more advanced level of science. Disregarding some technical logical questions, the data upon which the construction is based turn out to be items of immediate experience (sentience) and are thus "mental" after all, in one of the two senses of "mental" which we have been at pains to explicate.

This is not the place to review the many arguments * which have been advanced in the refutation of phenomenalism. If an epistemology with a phenomenal basis can at all be worked out satisfactorily, then these data have to be conceived as *lawfully* related to the physical objects of everyday life. This means that the doctrine of logical constructionism or reductionism, i.e., of the *explicit* definability of physical concepts in terms of phenomenal concepts, has to be abandoned. The logical relations involved here are *synthetic*, and the translatability thesis is not just utopian (owing to the always admitted complexities), but completely inadequate, if not quixotic. I remain unimpressed with the significance of Craig's theorem (cf. Hempel's essay in *Minnesota Studies in the Philosophy of Science*, Vol. II) in this connection. An infinite set of postulates is *not* what phenomenalists ever had in mind. And I believe there are other grave objections to that sort of a translatability doctrine. The kind of translatability which Craig's theorem allows for concerns only the empirical content of theories in the sense of all conceivable evidential (confirming) statements, but not in the sense of the *factual reference* of the postulates (and, hence, of the theorems).

Mutatis mutandis, it is now realized in many philosophical and psychological quarters † that the thesis of the translatability of statements about mental states (in phenomenal language) into statements about peripheral behavior (in *descriptive*, not theoretical physical_1 language) must also be repudiated.

With this firmly established orientation, the inference of sentience (raw feels) in other organisms seems prima facie restored to its original form as an argument from analogy. I have no doubt that *analogy* is the essential criterion for the ascription of sentience. But a closer look at the logic of the inference will prove worthwhile. The inference from peripheral behavior to central processes, very much like the inference from skulls to brains contained in them, is intersubjectively confirmable, and this in the sense that independent intersubjective evidence for the truth of these conclusions is in principle available. Just this is, of course,

* Cf. Freytag (128); Külpe (191); Broad (50, 51); Schlick (298); Reichenbach (273); Pap (248); Lovejoy (204); R. W. Sellars (307); W. S. Sellars (308); B. Russell (288); Kneale (179); Beck (24); Feigl (110, 111); Berlin (35); Watling (342); Braithwaite (48); E. J. Nelson (237, 238); and now, after a drastic change in outlook, even Ayer (18) is close to a critical realist position.

† Cf. Hempel (146); Carnap (64, 67, 73); Kaufmann (175); Jacobs (163); Pap (242, 243, 245, 248); Ayer (18); Feigl (113, 116); Cronbach and Meehl (79); Scriven (306).

not the case for the conclusions regarding mental states, if by mental states (sentience, raw feels) one means something that is not identifiable (i.e., not explicitly definable in $physical_1$ terms) with either overt-behavioral or central-neural states or processes.* If, contrary to the suggested orientation, such identifications could be made, i.e., if explicit definition could plausibly be given as an analysis of the meaning of phenomenal terms, then indeed no analogical inferences would be required. Nevertheless, considerations of analogy would be *suggestive*, though never decisive, for the *terminological* conventions according to which we apply or refrain from applying phenomenal terms to the behavior of animals and plants (let alone lifeless things).

If, however, phenomenal terms are logically irreducible to physicalistic terms, then parallelistic (epiphenomenalist) dualism is the most plausible alternative view. But *interactionistic* dualism is empirically much less defensible, and its methodological orientation too defeatist, to be acceptable to the current scientific outlook (cf. section II, above). And epiphenomenalism also has generally been considered objectionable because it denies the causal efficacy of raw feels; and because it introduces peculiar lawlike relations between cerebral events and mental events. These correlation laws are utterly different from any other laws of ($physical_2$) science in that, first, they are nomological "danglers," i.e., relations which connect intersubjectively confirmable events with events which *ex hypothesi* are in principle not intersubjectively and independently confirmable. Hence, the presence or absence of phenomenal data is not a difference that could conceivably make a difference in the confirmatory $physical_1$-*observational* evidence, i.e., in the publicly observable behavior, or for that matter in the neural processes observed or inferred by the neurophysiologists. And second, these correlation laws would, unlike other correlation laws in the natural sciences, be (again *ex hypothesi*) absolutely underivable from the premises of even the most inclusive and enriched set of postulates of any future theoretical physics or biology.

No wonder then that after a period of acquiescence with epiphenomenalism during the last century (T. H. Huxley, *et al.*), the behaviorist

* This is my way of stating succinctly the puzzle of "Other Minds" as it is understood in the long (unfinished) sequence of agonizing articles by John Wisdom (354), and in many other authors' publications, notably: Carnap (61, 62); Schlick (299); Ayer (15, 18); Austin (10); Pap (243, 248); Hampshire (141); Watling (341); Mellor (223).

movement in psychology took hold, and exercised an unprecedented influence in so many quarters. Behaviorists, in *their* way, repressed the problem either in that they *denied* the existence of raw feels (materialism); or in that they *defined* them in physical$_1$-observation terms (logical behaviorism); or they maintained that the subject matter of scientific and experimental psychology can be nothing but behavior (methodological behaviorism), which leaves the existence of raw feels an open question, but as of no relevance to science. Our previous discussions have, I trust, clearly indicated that behaviorism in the first sense is absurdly false; in the second sense it is inadequate as a logical analysis of the meaning of phenomenal terms; and in the third sense, it is an admittedly fruitful but limited program of research, but it entails no conclusion directly relevant to the central philosophical issue.

The repudiation of radical behaviorism and of logical behaviorism entails the acceptance of some sort of parallelistic doctrine. Recent arguments for this position * are prima facie highly persuasive. The basic point is simply that each of us knows his own states of immediate experience by acquaintance, and that by analogical reasoning we can infer similar, though never directly inspectable, states of experience in others. Direct inspection of the mental states of others is now generally considered a *logical* impossibility. For example, the subjunctive conditional, "If I were you, I would experience your pain," is not merely counterfactual, but counterlogical in that the antecedent of the conditional involves an outright inconsistency. The air of plausibility of the mentioned subjunctive conditional derives from entirely other, quite legitimate types of subjunctive conditionals, such as "If I had a broken leg (as you do), I should feel pain"; or "If I had (some traits of) your personality, flattery would please me." The logical grammar of personal proper names (or pronouns) however is such that it is downright self-contradictory to say (in a reasonably constructed and interpreted language) that Smith is Jones, or that I am you. The Mont Blanc cannot conceivably be identical with Mt. Everest!

Indirect verification or confirmation of statements regarding the mental states of other persons is however clearly possible once we have established laws regarding the correlation of the Φ's with the ψ's for our own case. And as we have pointed out, these laws could in prin-

* By Pap (243); Hampshire (141); Watling (341); Ayer (18).

ciple be most directly established with the help of an autocerebroscope. On the level of common life, of course, the correlations between neural and mental states are totally unknown. But a great many behavioral *indicators* are constantly being used in the (probabilistic) ascription of mental states. Logical analysis (Carnap, 73; Scriven, 306; Feyerabend, 119; Watling, 342; Feigl, 110, 111, 112, 114) has, I think, quite convincingly demonstrated the need for distinguishing the evidential bases from the factual reference of concepts and statements. The behavioral indicators serve as evidential bases for the ascription of mental states. Only the person who *experiences* the mental state can *directly* verify its occurrence. But there is no reason whatever to assume that when A reports his mental state, and B talks about it on the basis of behavioral evidence (or, if this is feasible, on the basis of neurophysiological evidence), that what they are talking about is not the very same mental state. This is indeed the way in which ordinary communication is understood. For example, if the doctor tells me a moment before lancing my abscess, "This will hurt," it is I who can directly verify this prediction. Moreover, most of us have learned from childhood on how to conceal our thoughts, feelings, sentiments, how to dissimulate, play-act, etc. And so we can justifiably say that behavioral symptoms do *not reliably* indicate mental states. In the light of the basic principles of normal induction and analogy, involving symmetry considerations, solipsism (with its arbitrary asymmetries) must be regarded as an absurdly false, rather than as a meaningless doctrine.

If we had completely adequate and detailed knowledge of the *neural* processes in human brains, and the knowledge of the one-one, or at least one-many ψ-Φ correlation laws, then a description of a neural state would be completely reliable evidence (or a genuine criterion) for the occurrence of the corresponding mental state. If these central neural events are essential intermediate links in the causal chain which connects stimuli with responses, then these central states are (probabilistically) inferable from stimulus-response situations. In this respect they have a logical status similar to the mental states as they are inferred from behavior in everyday life, or as the basis of psychological test situations. One may therefore wonder whether two steps of inference are really needed for a full logical reconstruction of the scientific ascription of mental states to other persons; the first step being the one from overt behavior to central neural events, and the second step being the one from neural events to mental states. I shall return to this question in subsection *E*, where I shall discuss the arguments for and against the identification of raw feels with the denotata of certain theoretical physical$_1$ (or physical$_2$) concepts.

C. The Cognitive Roles of Acquaintance. Various meanings of "acquaintance" and of "knowledge by acquaintance" were sorted out in section IV A. Our present concern is with the roles of acquaintance and of knowledge by acquaintance in the enterprise of science, especially in psychology. The first question I wish to discuss concerns the cognitive "plus," i.e., the alleged advantages of knowledge by acquaintance over knowledge by description. We may ask, for example, what does the seeing man know that the congenitally blind man could not know. Or, to take two examples from Eddington (93, 94), What could a man know about the effects of jokes if he had no sense of humor? Could a Martian, entirely without sentiments of compassion or piety, know about what is going on during a commemoration of the armistice? For the sake of the argument, we assume complete physical (1 or 2) predictability and explainability of the behavior of humans equipped with vision, a sense of humor, and sentiments of piety. The Martian could then predict all responses, including the linguistic utterances of the earthlings in the situations which involve their visual perceptions, their laughter about jokes, or their (solemn) behavior at the commemoration. But *ex hypothesi*, the Martian would be lacking completely in the sort of *imagery* and *empathy* which depends on familiarity (direct acquaintance) with the kinds of *qualia* to be imaged or empathized.

As we have pointed out before, "knowledge *of*," i.e., "acquaintance with," qualia is not a necessary condition for "knowledge *about*" (or knowledge by inference of) those qualia. A psychiatrist may know a great deal *about* extreme states of manic euphoria or of abject melancholic depression, without ever having experienced anything anywhere near them himself. In this case, of course, it must be admitted that the psychiatrist can get an "idea" of these extreme conditions by imaginative extrapolation from the milder spells of elation or depression which he, along with all human beings, does know by acquaintance. But the case is different for observers who are congenitally deprived of acquaintance with an entire modality of direct experience. This is the case of the congenitally blind or deaf, or that of our fancied Martian who has no emotions or sentiments of any kind. But I think it is also the case of human beings endowed with the entire repertory of normal sensory and emotional experience, when they introduce theoretical concepts in their science, such as the electromagnetic or gravitational fields, electric currents, and nuclear forces. We are "acquainted" with the perceptible things, properties, and relations on the relevant evidential bases which suggest the introduction of these concepts into the system of science generally, or which justify their special application in particular instances of observation.

In the context of the present discussion it does not matter very much whether we use the narrower, philosophical notion of direct acquaintance (restricted to the qualia of raw feels) or the wider commonsense or physicalistic notion of acquaintance (which includes the directly observable properties and relations of the objects in our everyday life environment). I think it does make sense to say that we do not know by acquaintance the "nature" of electric currents or of the forces within the nuclei of atoms. And although the congenitally blind have no acquaintance with color qualities or visual shapes, they may nevertheless come to have knowledge by inference at least of the neural correlates among the processes in the occipital lobes of the brains of persons with eyesight. The "intrinsic nature" of those neural processes remains unknown by acquaintance to the blind scientist, just as the "intrinsic nature" of electric currents remains unknown to scientists who have eyesight, and who have seen electrical machines and wires, have been tickled or shocked by electric currents, have seen electric sparks, have felt the heat produced by electric currents, have read voltmeters and ammeters, have observed the chemical and magnetic effects of electric currents, etc.

I trust my readers will not charge me with obscurantist tendencies. I do not at all share the view (e.g., Bergson's) that *genuine* knowledge is to be found only in direct acquaintance or intuition. Bergson, in his *Introduction to Metaphysics*, paradoxically claimed that metaphysics—the intuitive knowledge of intrinsic reality—is "the science that dispenses with symbols altogether." I wish to assert, quite to the contrary, that genuine knowledge is *always symbolic*, be it knowledge by acquaintance as formulated in direct introspective report sentences, or be it knowledge by description as, e.g., in the hypotheses of modern nuclear physics. If we knew all *about* electricity, magnetism, nuclear forces, etc., i.e., if we had a complete set of laws concerning those matters—this would be all we could possibly wish to know within the scientific enterprise. Anything added to this by way of "acquaintance" would be cognitively irrelevant imagery. Such imagery might be welcome from a poetic or artistic point of view. It might occasionally be helpful heuristically or didactically, but even in this regard it amounts only to pictorial bywork, and is often dangerously misleading. "Thou shalt not make unto thee any graven image . . ."

Our world, being what it is, can of course be known by description, in any of its parts or aspects, only on the basis of a foothold somewhere in direct acquaintance. This, it seems to me, is one of the cornerstones of any empiricist epistemology, old or new. But the new empiricism of recent times has come to recognize that it matters little just which areas of acquaintance are available or actually utilized for the "triangulation" of facts or entities outside the scope of direct acquaintance. The congenitally blind-deaf person, I stress again, could in principle construct and confirm a complete system of the natural sciences (including astronomy!) and the social sciences (including the psychology of vision and hearing, as well as the psychology of art and music appreciation!). It should go without saying that such a person, like Helen Keller, would normally depend upon information received from persons endowed with visual and auditory perception.

But, supposing such a human being could survive a long time as a solitary observer and was equipped with supreme intelligence and ingenuity, then one can well conceive of various modern instruments and devices (involving photoelectric cells, amplifiers, electromagnetic indicators, etc.) he could invent which would serve him in the detection of the stars, the chemical constitution of various substances, the behavior of animals, and so on—all accessible to him ultimately through, e.g., tactual pointer readings of one sort or another. All this is merely a picturesque way of saying that the "nomological net," i.e., the system of scientific concepts and laws, may be "tacked down" in a variety of alternative ways, either in several sense modalities (as in the normal case), or even in only one of them. To be sure, "triangulation of entities in logical space" is much easier and much more secure in the normal case. But, as we have pointed out, normal, unaided perception by itself is also quite insufficient for the confirmation of our knowledge regarding radio waves, infrared, ultraviolet, gamma radiations, cosmic rays, the molecular and atomic structure of matter, the motions and other physical and chemical characteristics of stars and galaxies, etc. Intricate instruments and ingenious theoretical constructions are indispensable in the case of normal (multimodal) perception as well. The difference between persons equipped with all normal sense organs and the deaf-blind is only one of degree, or of the speed with which they would, respectively, attain knowledge about the world in which they are embedded and of which they are parts.

Similar considerations apply to the advantages held by fully equipped persons in regard to psychological and linguistic or descriptive-semantical knowledge. If I have been trained by normal education to apply phenomenal terms (like "red", "green", "lilac fragrance", "rose fragrance", "sweet", "sour", etc.) to qualia of my own direct experience, then I can predict much more readily the application of these terms by other persons in the presence of certain specifiable visual, olfactory or gustatory stimuli. But predictions of this sort are based upon analogical inference; and they are in principle dispensable, because the discriminatory and verbal behavior of other persons is open to intersubjective test. Moreover, if we had a complete neurophysiological explanation of discriminatory and verbal responses we could derive these responses from the cerebral states which initiate them, and which, in turn were engendered by sensory stimulation. Analogously, whatever reliability empathetic understanding in common life, or "clinical intuition" in the psychologist's practice, may have is ultimately to be appraised by intersubjective tests. But the speed with which empathy or intuition do their work depends upon the breadth and the richness of the "experience" of the judge. It also depends upon his use of critical controls.

If the psychologist's personality type is radically different from that of his subject, he will have to correct (often to the point of complete reversal) his first intuitions. For example, an extremely extrovert person will find it difficult to "understand" an extreme introvert, and vice versa. If, however, the personalities are very similar, intuition may "click" readily, and it may even be frequently quite correct. The role of direct acquaintance in all these cases simply amounts to having in one's own experience features and regularities with which one is quite familiar, and which are hence speedily projected and utilized in the interpretation of the behavior of other persons. I conclude that the advantages of direct acquaintance pertain to the *context of discovery* (cf. Reichenbach, 273) and not to the context of justification. All the examples discussed do not differ in principle from the obvious examples of persons with "wide experience" as contrasted with persons with "narrow experience," in the most ordinary meaning of these terms. Someone thoroughly familiar with the weather patterns of Minnesota, or with the conduct of business in the Congress of the United States (to take two very different illustrations of the same point) will have the advantages of much speedier inferences and (usually) more reliable predictions than someone who has had no opportunity of long range observations in either case.

The philosophically intriguing questions regarding acquaintance are, I think, of a different sort. They are best expressed by asking, e.g., What is it that the blind man cannot know concerning color qualities? What is it that the (emotionless) Martian could not know about human feelings and sentiments? If we assume complete physical (i.e., at least $physical_1$) predictability of human behavior, i.e., as much predictability as the best developed physical science of the future could conceivably provide, then it is clear that the blind man or the Martian would lack only *acquaintance* and *knowledge by acquaintance* in certain areas of the realm of qualia. Lacking acquaintance means *not having* those experiential qualia; and the consequent lack of knowledge by acquaintance simply amounts to being unable to label the qualia with terms used previously by the subject (or by some other subject) when confronted with their occurrence in direct experience. Now, mere *having* or *living through* ("erleben") is not *knowledge* in any sense. "Knowledge by acquaintance," however, as we understand it here, is propositional, it does make truth claims; and although it is not infallible, it is under favorable circumstances so reliable that we rarely hesitate to call it "certain." It remains in any case the ultimate confirmation basis of *all* knowledge claims.

In many of the foregoing discussions we have suggested that what one person *has* and *knows by acquaintance* may be identical with what someone else knows by description. The color experiences of the man who can see are known to him by acquaintance, but the blind man *can* have inferential knowledge, or knowledge by description *about* those same experiences. After all, this is true as regards an individual color experience even if the other person *is* endowed with eyesight. The other person does not and could not conceivably *have* the numerically identical experience (see p. 30f above). Why should we then not conclude that the behavioristic psychologist can "triangulate" the direct experiences of others? I think that indeed he does just that if he relinquishes the narrow peripheralist position, i.e., if he allows himself the introduction of theoretical concepts which are only logically connected with, but never explicitly definable in terms of, concepts pertaining to overt molar behavior. These acquaintancewise possibly unknown states which the behaviorist must introduce for the sake of a theoretical explanation of overt behavior, and to which he (no longer a "radical" behaviorist) refers as the central causes of the peripheral behavior symptoms and manifestations, may well be *identical* with the referents of the phenomenal (acquaintance) terms used by his subject in introspective descriptions of his (the subject's) direct experience. As remarked before, in ordinary communication about our respective mental states, we make this assumption of identity quite unquestioningly. It took a great deal of training in philosophical doubt for learned men to call this assumption into question.

But philosophical doubt, here as elsewhere,* while stimulating in the search for clarity, is ultimately due to conceptual confusions. We have learned how to avoid these confusions, and thus to return with a good philosophical conscience to (at least *some* of) the convictions of commonsense. We have learned that philosophical doubts, unlike ordinary empirical doubts, cannot be removed by logical or experimental demonstration. What *can* be demonstrated logically is only the exploitation of certain misleading extensions of, or deviations from, the sensible and fruitful use of terms in ordinary or scientific language. Thus to doubt whether we can at all have knowledge about the "private" experience of other persons is merely the philosophical extension of the ordinary and quite legitimate doubts that we may have in specific instances, for example, when we ask "Is he really as disappointed as his behavior would seem to indicate?" This is to confuse practical difficulties of knowing with (allegedly) basic impossibilities. Once one becomes fully aware of the disease of philosophical skepticism, it becomes possible to cure oneself of it by a sort of self-analysis (*logical* analysis is what I have in mind here; but in certain cases psychoanalysis may help too, or may even be indispensable).

Granting then that the *referents* of acquaintance terms and $physical_1$ theoretical terms may in some cases be identical, this does not by itself

* As, e.g., in the problems of induction, the trustworthiness of memory, the veridicality of perception, etc.

decide the issue between monism and dualism. As we have seen in the previous subsection, the inference to other persons' raw feels can be *logically* differentiated from the inference to their central nervous processes. Dualistic parallelism or epiphenomenalism is entirely compatible with the assertion of the identity of the subjectively labeled mental state with the intersubjectively inferred state which is needed for the explanation of molar behavior. The mental state is logically distinguishable from the "correlated" neurophysiological state. Indeed (as pointed out in section III 4), it makes no sense to talk of *correlation,* or in any case not the usual sense, if the relation of "correlation" were that of *identity.* We shall tackle this crucial point in the next two subsections.

Before we proceed to the discussion of identity and identification, let us however summarize some important conclusions from our discussion of *acquaintance.* The data of direct experience function in three roles: First, in the use of typical patterns and regularities of one person's data for the intuitive or empathetic ascription of similar patterns and regularities of direct experience (or even of unconscious processes) to other persons, these data *suggest,* but by themselves are never a sufficiently strong *basis of validation* for knowledge claims about the mental life of other persons. Further clinical, experimental, or statistical studies of the behavior of those persons are needed in order to obtain a scientifically respectable degree of confirmation for such inferences. Second, nevertheless, and this is philosophically even more important, the first-person data of direct experience are, in the ultimate epistemological analysis, the *confirmation basis* of *all* types of factual knowledge claims. This is simply the core of the empiricist thesis over again. But third, the data are *also objects* (targets, referents) of *some* knowledge claims, viz. of those statements which concern nothing but the occurrence of raw feels or whatever regularities (if any!) can be formulated about raw feels in purely phenomenal terms. For examples of the latter, I mention the three-dimensional ordering of color qualia according to hue, brightness, and saturation; the regularities regarding the gradual (temporal) fading of intense emotions like joy, rage, exultation, embarrassment, regret, grief, etc.; the lawful correlations between, e.g., the experienced contents of daydreams and the attendant emotions of hope or fear. In all these cases, no matter whether the raw feels are our own or someone else's, *they* are the *objects* of our knowledge claims or the *referents* of certain terms in the sentences which describe them. I emphasize this point because recent empiricist epistemologies in their concern with the *confirmation* bases of our knowledge claims, and with observation statements which formulate the confirming (or disconfirming) evidence, have tended to neglect consideration of those cases in which the *target* of the knowledge claim is a state or a regularity of direct experience. Evidence and reference coincide only in the case of statements about the immediate data of first-person experience. But they are clearly distinct in all other cases, such as those in which the object of reference is a state of affairs in the world outside the observer (or else anatomically physiologically *inside* his own skin), no matter whether it be the state of inorganic things, or processes in organisms. Even the direct experience of oneself at a time distinct from the present moment, and of course the direct experience of other organisms or persons are numerically distinct from the data of the confirming evidence. In short, the data of immediate experience function either as *verifiers* or as *referents* of knowledge claims.

D. Reduction and Identification in Scientific Theories. In order to decide whether the mental and the physical can in some sense be identified, it is indispensable to cast at least a brief glance at the logic of reduction and identification in the sciences, especially in physics, biology, and psychology. Although these reflections will not provide us with the complete solution of the problem, they will be helpful and suggestive.

It was pointed out and briefly discussed in section II that the advance of scientific theories consists essentially in the reduction of a variety of originally heterogeneous observable facts and regularities to a unitary set of explanatory concepts and postulates. Customarily it is said, for example, that visible light *is* electromagnetic radiation (within a certain interval of wave lengths); that table salt *is* NaCl; that magnetized iron *is* an aggregate of iron atoms with a characteristic spin of certain of their electrons; that the transmitters of hereditary traits are the genes in the chromosomes of the germ cells; that (at least) short range memory traces are reverberating circuits in cerebral cell assemblies, etc. The "is" and the "are" in these sentences represent identities. But these identities differ in their mode of certification from the analytic identities of pure logic and mathematics. For extremely simple illustrations consider the general theorem of set theory "$[S \vee T] = -[-S \cdot -T]$" and the specific arithmetical identity "$\sqrt{64} = 2^3$" which hold by virtue of presupposed definitions and the principles of logic or arithmetic.

But the identities established in the factual sciences are confirmed on the basis of empirical evidence. This is very like the empirically ascertainable identity of Shakespeare (or could it be Marlowe?) with the author of *Hamlet*, or the identity of the author of *Hamlet* with the author of *King Lear*. Of course there are also such empirically ascertainable identities as those of Tully and Cicero, of William Thompson and Lord Kelvin, or of the evening star and the morning star. In the examples just given we have (extensional) identities of individuals labeled or uniquely described in two or more ways. When it comes to properties (universals), the identity may be either intensional or extensional. An illustration of the first is, e.g., the identity of d♯ and e♭ in the well tempered scale of music. An illustration of the second is the identity of the chemical element with atomic number or nuclear charge 20 with calcium characterized as a constituent of limestone, of atomic weight 40, having a melting point of 810° C., a specific heat of 0.169 at 20° C., etc.

In the case of analytic identities of individuals or of properties we may speak of the synonymy of names or predicates, respectively. (This applies, of course, also to two-place, three-place, etc. predicates, i.e., to dyadic, triadic, etc. relations. Thus, e.g., "earlier than" is logically synonymous with "temporally precedent to" or with the converse of the relation "later than"). The identity of the class of rational animals with the class of featherless bipeds (disregarding plucked birds), or with the class of laughing animals (disregarding hyenas), is *extensional* and *empirical*. Of course, extensional identity, be it logically necessary or empirical, is implied by intensional identity, but not vice versa. There is no longer any reason to be puzzled about identity being a *relation*. The proper explication of identity consists simply in the recognition that one and the same individual (or universal) may be designated by different labels or described by different characterizations. This could (but need not) be formulated by saying that the relation of identity fully explicated amounts to a triadic relation between labels (L), or descriptions (D) and a referent (R). The following diagrams represent the simplest paradigmatic situations.

$$L_1 \longrightarrow R \longleftarrow L_2 \text{ or } L \longrightarrow R \longleftarrow D \text{ or } D_1 \longrightarrow R \longleftarrow D_2$$

Since I am not a nominalist, having remained unconvinced by the arguments of Quine, Goodman, and White (269, 242), I see no objection to introducing *universals* as referents of predicates or relations. And since I am not a Platonic realist either, I am quite willing to consider talk about universals as a convenient *façon de parler*, rather than as a matter of profound "*ontological*" significance. In my previous example I regarded "d♯" and "e♭" as different labels for the same *kind* of musical tone-as-heard. Similarly I see no reason whatever to deny that "calcium" and "element of atomic number 20" designate the same *kind* of substance. This amounts to saying that the identity of universals, if it is not based on the *logical* synonymy of intensions, can amount only to an *extensional* (in this case, empirical) *equivalence* of two classes.

Prima facie the identifications achieved by scientific laws and theories appear to be cases of co-extensiveness, i.e., of extensional equivalence. This is certainly the case with identifications based on empirical laws. A metal characterized in terms of its thermal conductivity may be identical with the metal characterized by its electric conductivity. The ascertainment of the identity, in this case, depends upon the validity of the Wiedemann-Franz law according to which there is a linear relationship between the two kinds of conductivity. Now, while I grant that the word "*identity*" has only one meaning, and this is the meaning defined by the (properly understood) Leibniz principle of *identitas indiscernibilium*, the modes of *ascertainment* of identity are for our purposes the essential consideration. I shall therefore take the terminological liberty of speaking of different kinds of identity, viz., (1) logical, (2) empirical; and under (2) I shall distinguish (a) accidental, (b) nomological, (c) theoretical identities. In more precise but also more cumbersome language this would amount to distinguishing the various modes of ascertainment of identity, or the types of validity that assertions of identity may have.

The identity of the class of rational animals with the class of featherless bipeds may be considered not only as logically contingent, but as empirically accidental; in the same sense as we consider it empirically accidental that the city which is the seat of the United States Government is identical with the city in which on January 17, 1956, at 11:00 a.m. the temperature was (say) 43° F., the barometric pressure 30 inches, and the relative humidity 89 per cent. The referent of these descriptions is the one city of Washington, D.C. This is identity of individuals. Nomological identities rest on empirical laws; theoretical identities depend upon the postulates and definitions of a scientific theory. Since all types of identity, except the logical, are established on the basis of empirical evidence, they must therefore be formulated in *synthetic* statements.

There is, however, the temptation to regard certain well established *theoretical* identities as analytic. For example, if "gas pressure" is *defined* in terms of the sum of the momenta delivered by the molecules of a gas to the walls of its container, then of course within the context of the kinetic theory of gases, the identification of pressure with the sum of the molecular momenta is *analytic*. But, as Ernest Nagel (230) has made clear, if we mean by "the pressure of a gas" that property of it which is measurable by manometers, and which has a variety of well-known lawful connections with the volume, the temperature, etc. of the gas, and thus "manifests" itself in a variety of ways, then clearly it was a *discovery*, yielding *new* information, that revealed to us the relation of gas pressure (the "macro"-concept) to certain aspects of molecular motion. This is clearly synthetic. The interesting point which makes it so tempting to view the relation as analytic is, however, worth a little discussion. It is not simply the much vaunted arbitrariness of definitions.* It is rather that the macro-properties and macro-regularities of gases can be *derived* † from the assumptions of the molecular-kinetic theory. A full fledged micro-theory of thermal conduction, convection, diffusion, etc. thus enables us, among other things, to derive the regularities of such indicating instruments as the manometers, thermometers, etc. The expansion of the volume of the gas in the gas thermometer is an immediate logical consequence of the (assumed) increase in the average velocities of the molecules making up the gas, and the initial and boundary conditions which characterize the micro-state of the instrument. Quite analogous considerations apply to the electron theory of electric currents and the measurements of electromotive force and current intensity with the help of such indicating instruments as the voltmeter and the ammeter.

* What is arbitrary in definitions is usually very uninteresting and inconsequential, in contrast to what is *not* arbitrary.

† It was customary to assume that these derivations are *deductive*. But some of the premises in this case are *statistical* laws; hence some of the derivations of descriptive-observational or empirical-regularity conclusions are probabilistic. Strict deductions, however, can be found in *classical* thermodynamics, *classical* electrodynamics, in the theory of relativity and other examples of "classical" scientific theories. Even in statistical mechanics some derivations are strictly deductive, others so *highly* probable that for practical purposes they can be considered as ("nearly") deductive.

The explanation of the macro-behavior of organisms is sought along methodologically similar lines. Neurophysiological laws and neural-endocrine-muscular, etc. states will presumably suffice for the explanation of even as complex and intricate behavior as that of human beings. Disregarding the ultimately (possibly inevitable) *statistical* aspects of some of the laws or of the assumptions about initial and boundary conditions, the neurophysiology of the future (3000 A.D.?) should provide complete deductive derivations of the behavior symptoms of various central states whose ψ-correlates are the familiar sensations, perceptions, thoughts, beliefs, desires, volitions, emotions, and sentiments (known by acquaintance and described in phenomenal language). Perhaps I should make clear that I am here trying not so much to convince my readers of the feasibility of what he may consider an entirely utopian and quixotic program for science. I am rather concerned to argue *conditionally*, i.e., *if* this physicalistic program can be carried out, then there would be something like an *empirical* identification of the referents of molar behavior theory concepts with the referents of some neurophysiological concepts. In its logical and methodological aspects this would be quite analogous to the identification of, e.g., the property of magnetism (as conceived in the *macro*-theories of physics) with certain micro-structures and processes involving electron spins, etc., ascribed to the atom and quantum dynamics of ferromagnetic substances. These identifications, like all others of a similar kind * appear as analytic only because of the mentioned relations of deducibility which we know (or believe) to hold between the micro-theoretical and macro-nomological or macro-descriptive propositions.

But a more accurate analysis reveals invariably a synthetic-empirical feature *somewhere* in the context of such scientific explanations. Just *where* this feature is located depends largely on the nature of the logical reconstruction by means of which we analyze those explanations. In the case of the length of the mercury column in a thermometer, or the volume of the gas in a gas thermometer, the derivation of their (respective) expansions under the condition of increasing heat intensity is so direct that the "identity" appears deceptively as a *logical* one. But even here, empirical regularities enter in. In addition to considerations of the

* E.g., table salt = NaCl; Units of heredity = Genes; Light = electromagnetic waves; the chemical bond = electromagnetic forces playing between the atoms within a molecule; memory traces = reverberating neural circuits; etc., etc.

respective thermal expansion coefficients of gases or mercury as compared with those of the glass of the instruments, there are the laws of geometrical optics regarding the paths of the light rays, and the laws of psychophysics and of psychophysiology concerning the visual perception of the mercury column or of the indicator (e.g., a drop of ink) of the gas thermometer.

Just *where* we decidc to put the boundary (or "partition") between the data of observation and the inferred state of affairs is thus a matter of convenience in epistemological reconstruction. But somewhere we must put it, if we are not to lose sight of the *empirical* character of the relation between the *data* and the *illata*. In one reconstruction the data statements concern the observables of common life. This is the epistemology favored by thinkers like Popper, Carnap, Reichenbach, Hempel, Ryle, Black, Skinner, and W. Sellars. They all agree in *this* respect even if they differ sharply in others. They all accept in one way or another an intersubjectvie (physicalistic) thing-language as the basis of epistemological reconstruction. Bertrand Russell, in his later works, is about the only thinker who has made a valiant attempt to combine acceptance of a *phenomenal* basis with a *realistic* (non-phenomenalistic) reconstruction. This means that, as a realist, he has long ago abandoned the earlier phenomenalistic translatability doctrine, and has ever since regarded the relation between physical object statements and phenomenal data statements as one of probabilistic inference. I believe this position still needs considerable logical clarification, but I also believe that it is basically sound, in that it pursues the epistemological analysis down to data which involve only that minimum of inference which knowledge by acquaintance requires. (This was more fully discussed in the preceding subsection.)

No matter where the line is drawn between observables and inferred entities, the most adequate reconstruction, it seems to me, has to be rendered in any case in terms of nomological nets. To return to the temperature example, we may say that the intensity of heat in an oven is *indicated* by various observable effects, but is not identical with any single one of them, nor is it identifiable with a disjunction (or other logical function) of the observable indications. The intensity of heat is nomologically, and hence *synthetically*, related to the indications of indicators. This is not to be confused with the quite obviously synthetic character of the functional or statistical relations between the indications themselves. Empiricists, positivists, and operationists have of course always stressed the empirical character of *these* correlations.

But even when theories (spelling out nomological networks) are adumbrated only in the form of extremely vague "promissory notes," the practice of scientific thinking clearly demonstrates that theoretical concepts (hypothetical entities) are never reducible to, or identifiable with, observable data (or logical constructions thereof). When, e.g., the spirochaete *treponema pallida* was still undiscovered, the "disease entity" *general paresis* was conceived as the *causative* factor which "produces" the various symptoms of that disease. Examples of this sort could be multiplied indefinitely from all the sciences. Theoretical concepts are "anchored" in the observables, but are not logically (explicitly) definable in terms of the observables. To be sure, it is the "congruence," "consilience," "convergence," or whatever one wishes to call the testable correlations between the observables that allows for the introduction of fruitful theoretical concepts. It is indeed this consilience which provides the empirical basis for the specification of the meaning of theoretical concepts. Abstract postulates alone determine only their logical or mathematical structure, but never their *empirical* significance.

New evidential bases, such as the microscopic bacteriological findings, provide additional, and usually crucially important, "fixes" upon the theoretical concepts. Nevertheless they amount essentially to enrichments of the nomological net, and thus to a revision of the "weights" of the various other indicators. Thus, in present day pathology, the presence of the spirochaete is a *criterion* of general paresis, and even if many of the usual symptoms were absent, the disease would be ascribed to a patient if a sufficient concentration of the spirochaetes in the nerve tissues were verified. The fact that the bacteriological evidence is correlated with the (more "superficial") symptoms is of course something that only observations could have confirmed. But this need not prevent us from saying that the disease entity *general paresis* as construed *before*, or *independently* of, the evidence for the presence of the spirochaete, can be rightfully *identified* with the disease characterized with the help of the bacteriological evidence.

I conclude that it is proper to speak of "*identification*," not only in the purely formal sciences where identity consists in the *logical* synonymy of two or more expressions, but also in those cases in which the mode of ascertainment is empirical. The important consequence for our problem is then this: Concepts of molar behavior theory like habit strength, expectancy, drive, instinct, memory trace, repression, superego, etc., may yet be identified in a future psychophysiology with specific types of neural-structure-and-process-patterns. The identification, involving as it will, factual discoveries, is *empirical* in its mode of certification, but it is an identification nonetheless.

E. Arguments Concerning the Identification of Sentience with Neural Events. I shall now present, as explicitly as I can, the reasons for an *empirical* identification of raw feels with neural processes. I shall also discuss several apparently trenchant arguments that have been advanced against this identity theory of the mental and the physical. It will be advisable first to state my thesis quite succinctly, and to elaborate the arguments for and against it afterwards.

Taking into consideration everything we have said so far about the scientific and the philosophical aspects of the mind-body problem, the following view suggests itself: The raw feels of direct experience as we "have" them, are empirically identifiable with the referents of certain specifiable concepts of molar behavior theory, and these in turn (this was argued in the preceding subsection *D*) are empirically identifiable with the referents of some neurophysiological concepts. As we have pointed out, the word "mental" in present day psychology covers, however, not only the events and processes of direct experience (i.e., the raw feels), but also the unconscious events and processes, as well as the "intentional acts" of perception, introspective awareness, expectation, thought, belief, doubt, desire, volition, resolution, etc. I have argued above that since *intentionality* as such is to be analyzed on the one hand in terms of pure semantics (and thus falls under the category of the *logical*, rather than the psychological), it would be a category mistake of the most glaring sort to attempt a neurophysiological identification of this aspect of "mind." But since, on the other hand, intentional acts as occurrents in direct experience are introspectively or phenomenologically describable in something quite like raw-feel terms, a neural identification of *this* aspect of mind is prima facie not excluded on purely logical grounds. Unconscious processes, such as those described in psychoanalytic theory, are methodologically on a par with the concepts of molar behavior theories (as, e.g., instinct, habit strength, expectancy, drive, etc.) and hence offer in principle no greater difficulties for neurophysiological identification than the concepts of molar behavior theory which refer to *conscious* events or processes (e.g., directly experienced sensations, thoughts, feelings, emotions, etc.). As we have repeatedly pointed out, the crux of the mind-body problem consists in the interpretation of the relation between raw feels and the neural processes. The questions to be discussed are therefore these:

1. What does the identity thesis assert about the relation of raw feels to neural events?
2. What is the difference, if there is a difference, between psychophysiological parallelism (or epiphenomenalism) and the identity thesis?
3. Can the identity thesis be defended against empirical arguments which support an interactionistic dualism?
4. Can the identity thesis be defended against philosophical arguments which support dualism on the grounds of the alleged fundamental differences between the properties of direct experience and the features of physical (neurophysiological) processes?

Since I have already paved the way for at least partial replies to question 3, and to some extent also to 4, I shall now primarily concentrate on questions 1 and 2, and discuss the other issues more briefly whenever they will be relevant.

The identity thesis which I wish to clarify and to defend asserts that the states of direct experience which conscious human beings "live through," and those which we confidently ascribe to some of the higher animals, are identical with certain (presumably configurational) aspects of the neural processes in those organisms. To put the same idea in the terminology explained previously, we may say, what is *had-in-experience*, and (in the case of human beings) *knowable by acquaintance*, is identical with the object of *knowledge by description* provided first by molar behavior theory and this is in turn identical with what the science of neurophysiology *describes* (or, rather, will describe when sufficient progress has been achieved) as processes in the central nervous system, perhaps especially in the cerebral cortex. In its basic core this is the "double knowledge" theory held by many modern monistic critical realists.*

* Especially Alois Riehl, Moritz Schlick, Richard Gätschenberger, H. Reichenbach, Günther Jacoby, Bertrand Russell, Roy W. Sellars, Durant Drake, and C. A. Strong. To be sure, there are very significant differences among these thinkers. Russell has never quite freed himself from the neutral monism (phenomenalism) of his earlier neorealistic phase. R. W. Sellars and, following him on a higher level of logical sophistication, his son, Wilfrid, have combined their realistic, double-knowledge view with a doctrine of evolutionary emergence. Opposing the emergence view, Strong and Drake, originally influenced by F. Paulsen, adopted a panpsychistic metaphysics. My own view is a development in more modern terms of the epistemological outlook common

This view does not have the disadvantages of the Spinozistic doctrine of the unknown or unknowable *third* of which the mental and the physical are aspects. The "mental" states or events (in the sense of raw feels) are the referents (the denotata) of the phenomenal terms of the language of introspection, as well as of certain terms of the neurophysiological language. For this reason I have in previous publications called my view a "double-language theory." But, as I have explained above, this way of phrasing it is possibly misleading in that it suggests a purely analytic (logical) translatability between the statements in the two languages. It may therefore be wiser to speak instead of *twofold* access or *double knowledge*. The identification, I have emphasized, is to be *empirically* justified, and hence there can be no *logical* equivalence between the concepts (or statements) in the two languages.

On superficial reflection one may be tempted to regard the identification of phenomenal data with neurophysiological events as a case of the *theoretically* ascertainable identities of the natural sciences. "Theoretical identity" (explicated in section V *D*) means the sameness of the referent (universal or particular) of two or more *intersubjective* descriptions. For example, it is the atomic micro-structure of a crystal which is indicated ("described") by the optical refraction index, the dielectric constant, the magnetic permeability coefficient, and in greater detail evidenced by X-ray diffraction patterns. Similarly, the various behavioral indications for habit strength refer to a certain, as yet not fully specified, neurophysiological structure in a brain, which may ultimately be certified by more direct histological evidence. Logical Behaviorism admits only intersubjectively confirmable statements and hence *defines* mentalistic (phenomenal) terms explicitly on the basis of molar behavioral theoretical concepts. Thus, to ascribe to a person the experience of, e.g., an after-image amounts, within the intersubjective frame of reference, to the ascription of a hypothetical construct (theoretical concept), anchored in observable stimulus and response variables. This

to Riehl, Schlick, Russell, and to some extent of that of the erratic but brilliant Gätschenberger. The French philosopher Raymond Ruyer (289, 290) especially before he turned to a speculative and questionable neovitalism (293) held a similar view. Among psychologists W. Köhler (182, 183), E. G. Boring (40), and D. K. Adams (1), again differing in many important respects, hold similar monistic positions. Personally, I consider sections 22–35 in Schlick (298) as the first genuinely perspicacious, lucid and convincing formulation of the realistic-monistic point of view here defended. It is to be hoped that an English translation of this classic in modern epistemology will eventually become available.

theoretical concept may then later be identified, i.e., come to be regarded as *empirically* co-referential with the more detailed and deductively more powerful neurophysiological concept.

The empirical character of the identification rests upon the extensional equivalences, or extensional implications, which hold between statements about the behavioral and the neurophysiological evidence. In our example this means that all persons to whom we ascribe an after-image, as evidenced by certain stimulus and response conditions, also have cerebral processes of a certain kind, and vice versa. In view of the uncertainties and inaccuracies of our experimental techniques we can at present, of course, assert only a statistical correlation between the two domains of evidence. That is to say, the equivalences or implications are, practically speaking, only probabilistic. But in any case, the correlations as well as the theoretical identification of the referents indicated by various items of evidence are formulated in *intersubjectively* confirmable statements.

The identification of raw feels with neural states, however, crosses what in metaphysical phraseology is sometimes called an "ontological barrier." It connects the "subjective" with the "intersubjective." It *identifies* the referents of subjective terms with the referents of certain objective terms. But in my view of the matter there is here no longer an unbridgeable gulf, and hence no occasion for metaphysical shudders. Taking into account the conclusions of the preceding analyses of "privacy", "acquaintance", "physical", and of "identification", private states known by direct acquaintance and referred to by phenomenal (subjective) terms *can* be described in a public (at least $physical_1$) language and may thus be empirically identifiable with the referents of certain neurophysiological terms. Privacy is capable of public (intersubjective) description, and the objects of intersubjective science can be evidenced by data of private experience.

The application of phenomenal terms in statements of knowledge by acquaintance is *direct*, and therefore the verification of such statements (about the present moment of subjective experience) is likewise immediate. Phenomenal terms applied to other persons or organisms are used *indirectly*, and the confirmation of statements containing phenomenal terms (thus used) is *mediated* by rules of inference, utilizing various strands in the nomological net as rules of inference. Judging by the structure of one's own experience, there seems to be no reason to assume the existence of *absolutely* private mental states; i.e., there are presumably no "captive minds" in our world. This is of course a basic ontological feature of nature as we have come to conceive it. It is an *empirical* feature of a very fundamental kind, similar in its "basic frame" character to the $3 + 1$ dimensionality of space-time, or to the causal order of the universe. Such frame principles do not differ in kind, although they differ in degree of generality, from the postulates of scientific theories. Their adoption is essentially regulated by the rules of the hypothetico-deductive method.

Logical empiricism as it has come to be formulated in recent years (Carnap, 70, 73; Feigl, 116) recognizes the difference between direct observation (knowledge-by-acquaintance) statements and inferential statements as a *contextual* difference between direct and indirect confirmation. It does not matter precisely where, in our epistemological reconstruction, we draw the line between the observable and the inferred entities. But wherever we do draw it, the scope of the directly experienceable or of the directly observable depends on the identity of the experiencing and/or observing subject.* What is directly verifiable for one subject is only indirectly confirmable for another. And these very statements (expressed in the preceding two sentences) may be formalized in a pragmatic, intersubjective metalanguage.

Having formulated and in outline explicated the identity thesis, we now have to attend to several important points of philosophical interpretation. I reject the (Spinozistic) double aspect theory because it involves the assumption of an unknown, if not unknowable, neutral ("third") substance or reality-in-itself of which the mental (sentience) and the physical (appearance, properties, structure, etc.) are complementary aspects. If the neutral third is conceived as unknown, then it can be excluded by the principle of parsimony which is an essential ingredient of the normal hypothetico-deductive method of theory construction. If it is defined as *in principle* unknowable, then it must be repudiated as factually meaningless on even the most liberally inter-

* As I understand Dewey and other pragmatists, as well as contextualists like S. C. Pepper (254, 255), this point has been explicitly recognized by them. Cf. also the discussions by analytic philosophers, such as Hampshire (141), Watling (341), and Ayer (18). An exact logical account of the linguistic reflection of direct versus indirect verifiability has been given in the analysis of egocentric particulars (token-reflexive, indexical terms) by B. Russell (286), Reichenbach (274), Burks (58), W. Sellars (308, 312), and Bar-Hillel (20).

preted empiricist criterion of significance. But our view does not in the least suggest the need for a neutral third of any sort. This will now be shown more explicitly.

If a brain physiologist were equipped with the knowledge and devices that may be available a thousand years hence, and could investigate my brain processes and describe them in full detail, then he could formulate his findings in neurophysiological language, and might even be able to produce a complete microphysical account in terms of atomic and subatomic concepts. In our logical analysis of the meanings of the word "physical" we have argued that the physical sciences consist of knowledge-claims-*by-description*. That is to say that the objects (targets, referents) of such knowledge claims are "triangulated" on the basis of various areas of observational (sensory) evidence. What these objects are acquaintancewise is left completely open as long as we remain within the frame of *physical* concept formation and theory construction. But, since in point of empirical fact, *I* am directly acquainted with the qualia of my own immediate experience, I happen to know (by acquaintance) what the neurophysiologist refers to when he talks about certain configurational aspects of my cerebral processes.

There is danger at this point of lapsing into the fallacies of the well-known doctrine of structuralism, according to which physical knowledge concerns only the *form* or *structure* of the events of the universe, whereas acquaintance concerns the *contents* or *qualia* of existence.* This doctrine is to be repudiated on two counts. First, by failing to distinguish acquaintance (the mere *having* of data, or the capacity for imaging *some* of them) from *knowledge* by acquaintance (propositions, e.g., about similarities or dissimilarities, rank-orders, etc., of the qualia of the given), the doctrine fails to recognize that even introspective or phenomenological knowledge claims are *structural* in the very same sense in which *all* knowledge is structural, i.e., that it consists in the formulation of *relations* of one sort or another. Second, the realistic interpretation of physical knowledge which we have defended implies that whatever we "triangulate" from various bases of sensory observation is to be considered as "qualitative" in a generalized sense of this term. In the vast majority of cases the qualitative content of the referents of physical descriptions is *not* "given," i.e., it is not part of a phenomenal field.

* This doctrine has been espoused in various forms by Poincare (257), Eddington (93), C. I. Lewis (195), Schlick (299), *et al.*

But it *is* a given content in the case of certain specifiable neurophysiological processes.

If one wishes to trace the historical origins of this view, one might find it, if not in Aristotle, then certainly in Kant who came very close to saying that the experienced content is the *Ding-an-sich* which corresponds to the brain process as known in the spatio-temporal-causal concepts of natural science.* To put it more picturesquely, in the physical account of the universe as provided in the four-dimensional Minkowski diagram, there are sporadically some very small regions (representing the brains of living and awake organisms) which are "illuminated by the inner light" of direct experience or sentience. This view differs from panpsychism which assumes that the "internal illumination" pervades *all* of physical reality. But the panpsychists' hypothesis is inconsistent with the very principles of analogy which they claim to use as guides for their reasoning. If one really follows the analogies, then it stands to reason that the enormous differences in behavior (and neural processes) that exist between, e.g., human beings and insects, indicate equally great differences in their corresponding direct experience or sentience. Fancying the qualities of sentience of the lower animals is best left to poetic writers like Fechner, Bergson, or Maeterlinck. As regards the mental life of robots, or of Scriven's (304) "androids," I cannot believe that they could display *all* (or even most) of the characteristics of human behavior unless they were made of the proteins that constitute the nervous systems—and in that case they would present no puzzle.

The identity view here proposed has met with a great deal of resistance, especially on the part of modern analytic philosophers. To be sure, there are identifications which are "above suspicion." For example, it has been suggested that a legitimate form of empirical identification is to be found in such paradigms as the identity of the "visual" with the "tactual" *penny* (or the visual, tactual, and olfactory *rose*; or the visual, tactual, and auditory *bell*). In each of these examples one may distinguish the various domains of sensory evidence from the particular thing (or thing-kind) that the evidence indicates or refers to. Phenomenalists will, of course, be quick to point out that there is no sense in talking of a thing existing over and above the actual and possible "evidential" data and their important correlations. But from my realistic

* Cf. I. Kant, *Critique of Pure Reason*, section on "The Paralogisms of Pure Reason."

point of view it makes perfectly good sense to explain in terms of physical, psychophysical, and psychophysiological theories how, e.g. a bell by reflecting light, producing sound waves and being a solid, hard body affects our retina, cochlea, and our tactile nerve endings (under specifiable perceptual conditions) and thus produces the visual, tactual, and auditory data in our direct experience. This is indeed the "causal theory of perception" so much maligned by phenomenalists.

We grant that as empiricists we must ultimately justify the causal theory of perception (which is indeed a *scientific* theory, and not an epistemological analysis) by reference to the evidential data which confirm it. And this we can do, no matter whether our own perceptions are concerned (in the egocentric perspective) or those of others (in the "side view" or lateral perspective that we obtain by observing the stimuli, central processes and responses pertaining to other persons). The various sensory "aspects" of the bell are thus to be conceived as the effects which the bell, considered either on the common sense level, or on the micro-level of scientific analysis, has upon our sense organs and finally on our awareness (this last effect empirically identifiable with processes in various cortical areas). Since the phenomenalist thesis of the translatability of physical object statements into data statements is untenable, epistemological analysis must "dovetail" with the causal (scientific) theory of perception and render justice to the latter by an explicit reconstruction of the *nomological* (not purely logical!) relations between the data and the illata. This is still *conceptual* analysis, in that it retraces the relations between the concepts of stimulus objects and the concepts pertaining to the central (cortical-mental) processes in the perceiving organisms.

Our ψ-Φ identification, however, *cannot* be conceived according to the paradigm of the identity of stimulus objects (like the bell, or the rose). The analogy is misleading in that we have, in the case of stimulus objects physical descriptions of them which together with the empirical laws of psychophysics and psychophysiology enable us (in principle) to derive their various sensory "appearances." Far from requiring an unknown or unknowable "third" or "neutral propertyless substance," ordinary knowledge and especially scientific theory contains a great deal of information about the nature and structure of stimulus objects. The situation in the ψ-Φ case is fundamentally different: We don't have two kinds of *evidence* for one and the same entity (event, process, etc.). In direct acquaintance we *have*, we *experience* the datum (it is not evidenced, it is *evident*!), and we identify it with a physical process which we posit as an illatum whose existence is asserted on the basis of multifarious data in other evidential domains.

It should now be clear how the view here proposed differs from the Spinozistic double aspect doctrine. The data of experience are the reality which a very narrow class of neurophysiological concepts denotes. I admit this sounds very "metaphysical." And I shall no doubt be accused of illegitimately extending the ordinary meaning of "denotation". I am fully aware that I am *extending* the meaning. But I plead that this does not involve my view in paradoxes or needless perplexities. It is true that in common parlance, as well as in the widely accepted philosophical usage, we would say that a term like "neural process in the occipital lobe" *denotes* a pattern of nerve currents, and not a visual experience. But this remark obviously comes down to the true but trivial semantical assertion that a term designates its designatum; (e.g., "neuron" designates neuron!).

A specification of meaning can be attained through semantic designation rules only if the meaning of the translation equivalent of the definiendum is already understood in the metalanguage. Obviously, according to the commonly accepted usage of the word, a "denotatum" is the referent of proper names, and (except for the null cases) also of predicates, relations, etc. A genuine specification of meaning for empirical terms can be achieved only by a combination of semantical, syntactical and pragmatic rules. The last two types of rules are particularly important. The syntactical rules specify the relations of concepts to one another, and the pragmatic ones make clear which concepts pertain to a basis of direct evidence. The realistic interpretation of empirical concepts depends on an appropriate analysis especially of the roles of proper names (and in scientific languages of coordinates) and of individual-variables (coordinate-variables).*

Taking these analyses into account, we can recognize the valid elements in the older critical realistic epistemology of perceptual and conceptual reference. A physical object or process as perceived in common life, or as conceived in science, is the referent of certain symbolic representations. I submit that it is the preoccupation with the confirmatory

* Cf. especially W. Sellars (308); H. Feigl (110, 111); Bar-Hillel (20).

evidence which has misled positivists and some pragmatists (all of them phenomenalists, radical empiricists, or operationists) to identify the meaning of physical object statements with the actual and/or possible data which, according to our view, merely constitute their evidential bases. Worse still, even sophisticated analytic philosophers tend to confuse the meaning of physical concepts with the perceived or imaged appearance of physical things. No wonder then that we are told that the identity of certain neurophysiological states (or features thereof) with raw feels is a logical blunder. If the denotatum of "brain process (of a specified sort)" is thus confused with the appearance of the gray mass of the brain as one perceives it when looking into an opened skull, then it is indeed logically impossible to identify this appearance with the raw feels, e.g., of greenness or of anxiety.

It would be a similarly bad logical blunder to identify such raw feels with the scientific (heuristic or didactic) tinkertoy models of complex molecular structures (as of amino acids, or proteins) displayed by chemistry instructors in their courses. I don't know whether I should call these blunders "category mistakes." The first one simply consists in the confusion of evidence with the evidenced, or of the indicator with the indicated. What mistake does one make if one confuses smoke with fire, footprints with a man walking, certain darkish spots on an X-ray photograph with tuberculosis? It is strange that of all people it should be the analytic philosophers (who would expose *these* fallacies with ruthless irony) who do not see that they are making the same sort of mistake in thinking that physical-object concepts denote the perceptual appearance of physical things.

As I have been at pains to point out (in section IV), the only consistent and philosophically fruitful meaning of "physical" (more precisely, of "physical_1") is that of a conceptual system anchored in sensory observation and designed for increasingly comprehensive and coherent explanations of the intersubjectively confirmable facts of observation. This conceptual system or any part of it is in principle non-intuitive (*unanschaulich* as the Germans call it, i.e., unvisualizable). Hence, an identification of a small subset of its referents with something directly given and knowable by acquaintance is in principle left completely open. In point of fact, the imagery commonly, and sometimes helpfully, employed in the thinking of theoretical physicists, biologists, or neurophysiologists consists primarily of *pictorial appeals*. These are at best intellectual crutches, fruitful only heuristically or didactically, and not to be confused with conceptual meanings. The fallacy of "introjection" * which was so vigorously criticized by Avenarius (the empiriocriticist of the last century) consists in the *pictorial* ascription of raw feels to other organisms. As we have seen, such ascriptions indeed clash with the (equally pictorial) ascriptions of physical-appearance properties to other persons or animals.

In the perceptual awareness of other organisms we are confronted with their *behavior*, i.e., their responses, facial expressions, tone of voice, gait, posture, linguistic utterances, etc., but never with their raw feels. Raw feels do not and cannot be fitted into the appearance picture. They must therefore be conceived as the subjective counterpart of these appearances. As such they are inferentially attainable but not perceptually accessible. At an earlier point we have already discussed the phenomenology of the alleged intuitive or empathetic apprehension of the mental states of other organisms. Since we must recognize intuitive or empathetic ascriptions as fallible and corrigible, they have to be regarded as *inferential* from the point of view of *logical* reconstruction (i.e., in the context of justification), no matter how immediate, "self-evident," compelling, or convincing they may be *psychologically*.

That "introjection" in this sense leads to absurdities becomes especially clear when we consider the ascription of phenomenal fields, e.g., of visual spatiality to other persons. Unless we are solipsists, there is every good reason in the world to ascribe to others the same sort of "life space" (phenomenal environment) which we find so distinctly within our own experience. But if we think of other persons in terms of their appearance in our own phenomenal environment, then it is impossible to ascribe (pictorially) to them also the particular perspectives that they perceive of their environment (or of parts of their own bodies). The fallacy is just as gross as in the case of expecting to find in the brain of another person looking at a green tree a little picture of that tree. But pictorial thinking is one thing, and conceptual thinking is quite another. For *conceptual* ascription, however, there is no difficulty. The *concepts* of neurophysiology are non-intuitive and must not be confused with their logically irrelevant pictorial connotations. These connotations lend, psychologically speaking, a certain "root flavor" to

* The term "introjection" as used by R. Avenarius has nothing to do with the well-known homonymous psychoanalytic concept.

these concepts. But once the pictorial appeals connected with the evidential roots of our physical or neurophysiological concepts are dismissed as irrelevant, they no longer pre-empt those places in the conceptual system of which we may then say that they denote some raw feels.

For these reasons I think that once the proper safeguards are applied, no category mistakes are made if we combine phenomenal and physical terms, as indeed we do quite ingenuously not only in ordinary discourse but also in the language of psychology. There is no reason why we should not say, e.g., "The anticipation of success quickened his pace"; "Morbid and tormenting thoughts caused his loss of appetite"; "Touching the hot stove caused intense pain"; "His repressed hostilities finally produced a gastric ulcer"; etc. Category mistakes do arise from confusions of universals with particulars; or of dispositions with occurrents. The first sort of category mistake certainly consists in a violation of the Russellian rule of types. I am not sure whether the second sort can always be reduced to the first. But the original diagnosis made especially by Carnap in his early (phenomenalistic) work (60) of the mind-body perplexities as Russellian-type confusions is no longer acceptable. Physical concepts are *not* logical constructions out of phenomenal concepts.

A more serious objection to identification comes from reflections upon Leibniz's principle of the identity of indiscernibles. Since we have not only admitted, but repeatedly emphasized the *empirical* nature of the ψ-Φ identification, one may well ask how we can speak of *identity* if its confirmation requires the observation of empirical regularities. The most direct confirmation conceivable would have to be executed with the help of an autocerebroscope. We may fancy a "compleat autocerebroscopist" who while introspectively attending to, e.g., his increasing feelings of anger (or love, hatred, embarrassment, exultation, or to the experience of a tune-as-heard, etc.) would simultaneously be observing a vastly magnified visual "picture" of his own cerebral nerve currents on a projection screen. (This piece of science fiction is conceived in analogy to the fluoroscope with the help of which a person may watch, e.g., his own heart action.) Along the lines of the proposed realistic interpretation he would take the shifting patterns visible on the screen as evidence for his own brain processes. Assuming the empirical core of parallelism or isomorphism, he would find that a "crescendo" in his anger—or in the melody he heard—would correspond to a "crescendo" in the "correlated" cortical processes. (Similarly for "accelerandos," "ritardandos," etc. Adrian's and McCulloch's experiments seem to have demonstrated a surprisingly simple isomorphism of the shapes of geometrical figures in the visual field with the patterns of raised electric potentials in the occipital lobe of the cortex.) According to the identity thesis the directly experienced qualia and configurations are the realities-in-themselves that are denoted by the neurophysiological descriptions. This identification of the denotata is therefore *empirical*, and the most direct evidence conceivably attainable would be that of the autocerebroscopically observable regularities.

Any detailed account of the ψ-Φ identities is a matter for the future progress of psychophysiological research. But in the light of the scanty knowledge available even today, it is plausible that only certain types of cerebral processes in some of their (probably configurational) aspects are identical with the experienced and acquaintancewise knowable raw feels. A "psychological physiology" * which frames hypotheses about neural structures and processes on the basis of a knowledge of the characteristics and the regularities in the changes of phenomenal fields must therefore always remain extremely sketchy. Knowledge by acquaintance of phenomenal fields alone cannot possibly yield more than a few strands of the total nomological net of neurophysiological concepts required for the explanation of molar behavior. The identification is therefore restricted to those elements, properties, or relations in the neural processes which (in dualistic parlance) are the "correlates" of the raw feels. In our monistic account this is tantamount to the identity of the denotata directly labeled by phenomenal terms, with the *denotata* of neural descriptions. These latter denotata are acquaintancewise unknown to the neurophysiologist, except if he uses the autocerebroscope himself.

Now it is clear that neural correlates (to speak for the sake of easier exposition once more dualistically) are denoted by concepts which are much richer in meaning than the corresponding phenomenal concepts. The neurophysiological concepts refer to complicated, highly ramified patterns of neuron discharges, whereas their raw-feel correlates may be simple qualities or relations in a phenomenal field. How can, e.g., a uniform patch of greenness, a single musical tone, a stinging pain be identical with a complex set of neural events? Here again it is essential to distinguish between the *scientific* and the *philosophical* components of

* Advocated by W. Köhler (184, 185) and critically discussed by C. C. Pratt (260).

this question. Our psychophysiological ignorance is still too great to permit anything more than bold guesses on the scientific side.

There has been talk of "thresholds" and "fusion"; i.e., it is assumed that raw feels emerge only if the intensities of the neural patterns have reached a certain degree; and that complex neural patterns may be "fused" so that the emerging quality "appears" simple and uniform. This sort of talk, though dangerously apt to mislead, is not entirely illegitimate. Talk of thresholds, limens, and fusion is of course quite customary and proper in *psychophysics*, but its extension to *psychophysiology* is precarious. It makes perfectly good sense, and is true, to say that the white and black sectors on a swiftly rotating disk phenomenally fuse and yield a uniformly gray appearance. It makes perfectly good sense also, and is equally true, that the intensity of physical stimuli (like light, sound, pressure on one's skin, concentration of chemical substances in the air, etc.) must surpass a certain lower limiting value, if they are to effect a sensation in any of the various modalities (sight, hearing, touch, smell, etc.).

If these facts have any analogies in the intra-cerebral sphere, it would have to be assumed that one area of the cortex "taps" or "scans" other areas and could thus not come to react unless the input reaches a certain intensity. Likewise, one would have to assume that the effect in the second area reflects only certain gross features of the intricate and multifarious process patterns in the first. These would be the analogues of psychophysical thresholds and fusions. Finally, one may assume that the second area (which corresponds to the sensing of the raw feels) is connected with another area corresponding respectively to awareness or judgment (as in introspection) and finally to a motoric area of the cortex which innervates expressive responses or speech.* May I say again that I don't for a moment insist on the scientific adequacy of this particular model. I am not trying to do armchair neurophysiology. All I am concerned to point out is that models are conceivable which would enable us to remove the obstacles arising from the apparent disparities of phenomenal unity versus physical multiplicity; phenomenal spatialities and physical space; phenomenal time and physical time; phenomenal purposiveness and physical causality; etc. I am now going to outline these considerations very briefly.

* I am indebted to R. Carnap for suggesting (in conversations) this sort of brain model.

W. Köhler (182, 183, 185) and R. Ruyer (290, 292, 293) have convincingly shown that the notorious Cartesian perplexities regarding spatiality can be removed by closer attention to the facts of psychophysiology combined with a logical clarification of the distinction between phenomenal space(s) and physical space. (We have laid the groundwork for this in section III *B*). The surface of objects "physically" outside my skin naturally appears in my visual space as external to the visual appearance of those parts of my body which I can see. There is histological and physiological evidence for a relatively simple projection of the excitation patterns in the retina of the eye, in the area of the occipital lobe of the cerebral cortex. The projection, in its physical and geometrical aspects, is similar to the sort of projection one gets on the screen of a periscope inside a submarine. Not only parts of the surrounding surface of the sea and of other ships, but also parts of the (surfacing) submarine itself are projected upon the screen. Similarly, when I lie on a couch I find not only the appearances of tables, chairs, walls, and windows within my visual field, but I find these object appearances phenomenally outside that part of my phenomenal body (chest, arms, hands, legs, feet) which is also included in my visual field. These simple reflections show that some of the older philosophical puzzles about the outward projection of visual percepts from my mind or brain into the external world are gratuitous, based on confusions, and resolvable by proper attention to the scientific facts on the one hand and to the meanings of spatial terms and phraseologies on the other.

The resolution of the perplexities regarding phenomenal versus physical time, as well as experienced purposiveness versus physical or physiological causality proceeds quite analogously. In the phenomenally temporal "projection" we locate ends-in-view at some distance in the future, and then go about attaining these ends by action, i.e. by the utilization of means. If, e.g., I decide to attend a lecture, I may have to go through a long chain of acts, such as walking to my garage, starting my car, driving to the auditorium, and getting seated there. My actions are clearly goal directed, but there is no need for the myth about the later events (the goal) influencing my antecedent behavior. My behavior is guided, controlled, or modulated by the goal *idea* which is contemporaneous with my instrumental acts, or possibly precedes them. What in the phenomenal description appears like a future event in my life career determining my current behavior, becomes in the causal account the effect of one part of my cerebral processes upon another. Of course in this case, just as in the case of memory (recollection), our thinking is essentially mediated by symbols; and therefore "intentionality" (cf. section IV *F*) plays an important role here. But the symbolic *representation* of past events or of future events is effected by processes occurring *now*; i.e., these representations are causal factors in the determination of current behavior. Just as there is no need for a curious notion of "final" causes (or, in Lecomte de Noüy's phrase, of "telefinality"), there is no need for the assumption of a literal presence of the past in present recollections. Whatever the adequate and detailed neurophysiological account of memory traces may ultimately turn out to be, it is these memory traces and not some direct and mysterious apprehension of past events which will causally account for the facts of recollection and of the modification of behavior through learning processes.

Similar considerations would seem to apply to the perennial puzzles concerned with the problems of the nature of the "*self*," i.e., the unity of the ego, or the unity of consciousness. Here, as in the other puzzles just discussed, the phenomenological descriptions may be correlated with the neurophysiological explanations. Phenomenally there may or may not be a "central core," the "I," in all my experiences. We may admit, following Hume and the later empiricists in the Humean tradition, that there is no distinct element, datum, or impression that could properly be regarded as the self. But it is hard to deny that in the directly given data and in their succession throughout experienced time, there is a certain feature of centralization, coordination, organization, or integration—the reader may choose whichever term seems most suitable. This unitary organization seems to rest on the ever-present potentialities of recollecting a great many events or sequences of events of one's (*sic*!) past; the ever present possibility of the occurrence of somatic data (referring to one's own body); the existence of a set of dispositions or behavior tendencies, including those ascribed (psychoanalytically) to the superego (i.e., in plain language our set of values and ideals as incorporated in one's conscience); and finally that conception of one's self which is largely a result of the realization of one's own character and personality, adequately or often very inadequately derived from interpretations of one's own behavior and one's social role as perceived by oneself or by others in the social context.

Whichever of these aspects are in some sense phenomenally "given"—

and I suggest a good many may well be so given—these aspects very likely "correspond" to (or according to my view, are identical with) certain relatively stable patterns of cerebral structures and functions. In the pathological cases of split or of alternating personalities (of the Sally Beauchamp, or of the Dr. Jekyll and Mr. Hyde varieties), it has often been suggested that we deal with cerebral subsystems, each having "organic unity" in itself, but only one of them dominating in the determination of behavior during certain intervals of time. If according to psychoanalytic theory large parts of the *id* as well as of the *superego* are unconscious, this may well be interpreted by assuming that certain portions of the cerebral processes are blocked off (this corresponds to "repressed") from the areas of awareness and of verbal report.

Having rendered plausible the *scientific* feasibility of at least a *parallelistic* account of some of the striking and remarkable features of mental life, I return now to the *philosophical* or *logical* crux of the *identity* thesis. We have stressed that the (empirical!) identification of the mental with the physical consists in regarding what is labeled in knowledge by acquaintance as a quale of direct experience as identical with the denotatum of some neurophysiological concept. The scientific evidence for parallelism or isomorphism is then *interpreted* as the *empirical* basis for the identification. The step from parallelism to the identity view is essentially a matter of philosophical interpretation. The principle of parsimony as it is employed in the sciences contributes only one reason in favor of monism. If isomorphism is admitted, the dualistic (parallelistic) position may be retained, but no good grounds can be adduced for such a duplication of realities, or even of "aspects" of reality. The principle of parsimony or of inductive (or hypothetico-deductive) simplicity does oppose the operationistic predilection for speaking of *two* (or more) concepts if the evidential facts, though completely correlated, are qualitatively heterogeneous.

Our view of "triangulation" under such conditions of convergence has, I trust, shown the operationist view to be by far too restrictive. But there is still the *logical* question how concepts with such fundamentally different evidential bases can be interpreted as (empirically) identifiable. In the case of the concept of the electric current (cf. above section V C) as measured by its magnetic, chemical or thermal, etc. effects, the identification of the several operationally introduced concepts is plausible enough. But, it will again be asked, how can we speak of identity in the entirely different psychophysiological case where one of the concepts is characterized by the direct applicability of subjective acquaintance terms and the other (the physiological) is introduced on an intersubjective basis and thus has its evidential roots in the sensory data of any qualified observer? I think the answer is not so difficult any more. If we first consider "acquaintance" in its *ordinary* usage, we can certainly say that Anthony Eden is *acquainted* with Queen Elizabeth II, and I am not (never having had the opportunity of meeting her). Nevertheless, I can lay claim to some knowledge about the Queen, based on newspaper reports, pictures, and the like. It is surely the *same* person that Eden and I know, each in his way. Closer to the point, I know by acquaintance what it is to have an eidetic musical-image experience (I occasionally "hear with my inner ear" entire passages from symphonies, string quartets, etc. in their full tone colors). Someone else lacking this sort of experience does not know it by acquaintance, but he can know *about* it, especially if he is a skillful experimental psychologist. It would be unparsimonious to assume that the psychologist and I are *referring* to two different (but correlated) processes.

Now, direct acquaintance with "private" raw feels is describable also in the intersubjective language of science. Its ultimate explanation may again have to refer to various cerebral areas, one of which (speaking for ease of exposition again dualistically) "corresponds" to *sensing*, another to *judging*, and possibly another yet corresponds to (introspective) *reporting*. I conclude that acquaintance statements differ only in the type and domain of *evidence*, but not in regard to their *reference*, from certain neurophysiological statements. Since the neural apparatus of introspection differs most markedly from that of (external) perception, it should not be surprising that knowledge by acquaintance (now taken in its *narrow* epistemological sense) is so much more crude, undetailed, and imprecise, than knowledge based on sense perception, especially when this is aided by the instruments of science.

Direct awareness, as we have pointed out before, usually furnishes only qualitative or topological orderings of the contents of phenomenal fields. It could not by itself inform us about the cerebral localization of subjective experience. A very crude (but, if taken literally, I fear highly misleading) analogy might help illuminate this point. A man lost in a jungle perceives the trees and undergrowth in his immediate environment. But the location of this very same part of the jungle can be determined in a much more accurate and encompassing manner by a cartographer making his measurements from the vantage point of an airplane or balloon high above the jungle. This simile is misleading, of course, in that *both* the lost wanderer *and* the cartographer use *sensory* perception as evidential bases for their knowledge claims. This clearly differs from the case in which I report (or "avow" as Ryle puts it), e.g., a feeling of anxiety and a behavioral psychologist infers my anxiety from the "symptoms," or a neurophysiologist recognizes it in the "corresponding" cerebral processes. Nevertheless, I fail to see that the difference, important though it is in many ways, affects the argument for the identification of the *referents* of the introspective avowal, with those of the two scientific descriptions.

I conclude that ψ-Φ identity as I conceive it is then still an identity of indiscernibles as defined by Leibniz and Russell. But as the clarification of the "paradox of analysis" (cf. Feyerabend, 120) and of related puzzles about belief sentences should by now have made amply clear, mutual substitutivity even of *logically* synonymous expressions holds only in non-pragmatic contexts. The *empirical* synonymy of ψ and Φ terms (or, more cautiously perhaps, their empirical *co-reference*) a fortiori does not allow for substitutivity in pragmatic contexts. By this I mean that the "*salva veritate*" condition is fulfilled only in contexts of substitution which do *not* depend on what we *know*, or what *evidence* we have for our knowledge claims. As we pointed out before, there are or were many people (primitive, ancient, etc.) who have no idea of the association of mental life with cerebral processes. But it is nevertheless as justifiable to speak of identity here as it is in the case of "Walter Scott = the author of the Waverley novels," regardless of whether this fact is known or unknown to a given person. In this particular and well-worn example the identity concerns an individual. But, not being a nominalist, I see no difficulties in the identity of a universal, named or described in various ways. Psychophysiological identity may be identity of particulars (*this* twinge of pain with a specific cerebral event at a certain time), or of universals (pain of a certain *kind*, and a *type* of cerebral process).

I am finally going to tackle more specifically and pointedly the question: What is the difference that makes a difference between the parallelism and the identity doctrines? The pragmatist-positivist flavor of this question suggests that it concerns empirically testable differences. But I have already admitted that there are no such differences and that there could not be any, as far as conceivable empirical evidence is concerned. Is the identity thesis then a piece of otiose metaphysics? Whether it is metaphysics depends of course on what one means by "metaphysics". As I see it, the question is not only similar, but indeed intimately related, to such "metaphysical" issues as realism versus phenomenalism, or the modality versus the regularity view of causality. As most philosophers nowadays realize, these issues unlike disputes regarding *scientific* theories cannot be decided by empirical tests. These questions concern the explication of the meaning of concepts and assumptions. They are a subject matter for logical analysis.

As to whether there is a tenable meaning of "causal necessity" related to regularity, but not reducible to it, this is a controversial issue today. My own reflections favor a view of causal modalities (possibility, necessity, impossibility) which explicates the use of these terms metalinguistically, and nevertheless does not conflict with Hume's basic, and in my opinion irrefutable, contention; viz., that (if I may put it in my own way) the only *evidence* we can ever have for the assertion of causal connections must be observed regularities. There is, as I see it, no *test* for causal necessity over and above the tests for regularity. But this does not preclude *meaning* from the distinction between *accidental* and *necessary* universal synthetic statements. A world is conceivable in which a certain metal with a high melting point (say, e.g., platinum) everywhere and always in the infinite history of that world occurs in the solid state, simply because the temperature in that world "happens" never anywhere to surpass a certain upper limit. In such a world the universal statement "$(x,y,z;t)(Pt_{xyzt} \supset S_{xyzt})$," i.e., "platinum is everywhere and always solid" would be a true universal statement. But the counterfactual conditional "if the temperature *were* ever to reach or surpass a certain value, platinum *would* melt" might even be deducible from the basic laws of physics of that world. The universal statement in question is accidentally true. It is not a consequence of a basic *law* of nature; its truth depends on certain contingent features of the initial and boundary conditions of the fancied world. This shows that there are meaningful distinctions for which no conceivable empirical test could be designed.

Even closer to our problem is the issue between realism and phenomenalism. As I have shown elsewhere (110), there is again no *testable* difference between these two interpretations of factual knowledge, but there are excellent reasons for the repudiation of phenomenalism and hence for the acceptance of a realistic epistemology. To relegate the issue to the limbo of metaphysics is a lazy man's way of saving himself the troubles of careful analysis. But close attention to the logic of evidence and reference shows that phenomenalism, even in its most liberal forms does not and cannot substantiate its translatability doctrine; and that only a view which relates phenomenal evidence *synthetically* to statements about physical objects is ultimately tenable.

It is precisely because realists locate both the *evidence* and the evidenced within the nomological net, that they can give a more adequate account of the relation between "the knower and the known" than positivists, pragmatists, or operationists have ever been able to provide. And it is for this very same reason, that our view of the nature of physical concepts enables us to identify *some* (of course very few only!) of their referents with the referents of raw feel terms. Dazzled by the admittedly tremendous importance of the evidential basis for our knowledge claims, positivists have regrettably neglected the very *objects* of those knowledge claims. They have myopically flattened them into the surface of evidence, and thus prevented themselves from giving a viable account of the concepts of physics; and they have merely evaded or repressed the mind-body problem which they thought would vanish if their "reductions"—phenomenalistic or behavioristic—were accepted. Ingenious and tempting though their more sophisticated endeavors of reduction have been, they did not succeed. This is why I felt that an explicit reinstatement and defense of a realistic solution of the mind-body problem would be timely and worthwhile.

VI. A Budget of Unsolved Problems. Suggestions for Further Analyses and Research

Although I have proposed what I believe to be at least a fairly circumspect sketch of an adequate solution of the mind-body problems, there are a number of specific component issues which require a great deal of further clarification and investigation. Since I am more interested in the continuing endeavors in this field than in having said the "last word" about it (that's almost inconceivable, in philosophy at any rate!), I shall now attempt to state and discuss succinctly a number of questions to which I have no entirely satisfactory answer at present. I should be immensely pleased if others were to take up these questions in their own work.

The foregoing analyses and discussions were intended to bring to a level of full awareness many of the repressed difficulties of our problem. I have been especially concerned to separate, as well as I could, the scientific from the philosophical issues. And I have tried to show that there are no *insuperable* logical difficulties for an identity theory of the mental and the physical. I shall again divide the discussion into two parts. The first (*A*, *B*, *C*) will be concerned with open *philosophical* questions and difficulties. The second (*D*) will appraise much more briefly the acceptability of identity theory in the light of possibly forthcoming heterodoxical *scientific* discoveries.

A. Is There a Phenomenal Language? The Relations of Meaning, Evidence, and Reference. The central core of the proposed solution rests upon the distinction between evidence and reference. No matter what indirect (behavioral) evidence we use for the ascription of mental states, the mental state ascribed is not to be confused with the evidence which only lends support to the ascription. A fortiori, we must eliminate the still worse confusion of the pictorial appeals (attached to evidential terms) with the conceptual meaning or the reference of neurophysiological concepts. The only case in which pictorial appeals or imagery may be thought to play an essential role in knowledge claims is at the ultimate phenomenal basis of the confirmation of all knowledge claims. And, as we have pointed out, if and only if these knowledge claims are so extremely restricted as to refer exclusively to a currently experienced datum, then—in this very special case—evidence and reference coincide. "Now green", "now anger", "now green spot on a gray background", "stinging pain suddenly increasing", etc. might be examples. The last example shows that the indexical term "now" need not appear in the phenomenal sentence; but of course the sentence is in the present tense, and this is presumably equivalent with the occurrence of the indexical "now".

It is difficult to decide whether indexical terms (i.e., egocentric particulars like "now", "I", "here", "this") are indispensable constituents of singular phenomenal sentences. There are, of course, many examples of *universal* statements which contain only *phenomenal* terms as descriptive signs (in addition to purely logical signs): "Orange is more similar to red than it is to green"; "Whatever is colored is extended (in the visual field)"; "Anger always subsides after some time"; etc. There is also the difficult question whether phenomenal sentences can contain proper names (or something like topological coordinates) for elements in the phenomenal fields. One of my examples suggested that one might use proper names for the small bright spots on the dark background of a visual field and thus describe their relative positions in terms of such relations as "to the left of", "above", and "far below". It seems clear that there is a danger of logical paradoxes, engendered by category mistakes, if we try to mix phenomenal sentences of this sort with the usual behaviorally based ascriptions of mental states to organisms. In these behavioral ascriptions the organism (or the person?) is the individual which is represented by the subject term of the sentence; the predicate is then something like "sees green", "sees an array of bright spots on a dark background". There can then be no direct translation of sentences in which the subject terms denote elements in a phenomenal field, into sentences in which the subject terms denote individual organisms. But perhaps there can be an *empirical* coreference between statements about some (configurational) aspects of neural fields and those about phenomenal fields.

The precise logical explication of empirical identity or coreference is fraught with many difficulties. Some of these stem from the tendency to think of *meaning* as *intension,* and then to conceive of intension in terms of its simplest picturable examples. *Blueness* is an intension indeed, but what are the intensions of "energy", "entropy", "electric field strength", "electric charge", "neuron discharge", "reverberating neural circuit"? In all these other cases the intensions are non-intuitive and can be specified only by postulates and correspondence rules. Similarly non-intuitive are the elements of the corresponding extensions, or the denotata. It does seem to me that we can rightly say that both the intension and the extension of the theoretical concepts of the physical sciences are largely unknown by acquaintance, and that only a very small selection of them can therefore be identified with the intensions and extensions of concepts-by-acquaintance. But of course the latter presuppose the existence of a phenomenal language. It has indeed been seriously questioned as to whether there is a phenomenal language at all. In the usual, and full-fledged sense, "language" means a symbolic system with specifiable syntactical (formation and transformation) rules, semantical (designation) rules, and pragmatic (verification) rules. Scraps and bits of phenomenal phraseology seem to fulfill these requirements, but an overall system like that of the physical language does not seem attainable.

The difficulties are further complicated by the question on which level of analysis we are to specify elements and relations described by phenomenal sentences. There is a long history of objections against the Hume-Mach-Russell-Price analysis of experience into "hard" and "soft" data. Phenomenologists, Gestalt psychologists, and more recently many analytic philosophers have raised serious objections not only against the atomism or elementarism of the sense-data doctrine, but also against any doctrine of immediacy or of the given.*

I have throughout this essay maintained and argued that genuinely phenomenal or acquaintance terms are indispensable, not only for the reconstruction of the indirect confirmation of practically all our knowledge claims, but also as labels for the referents of some knowledge claims—whether they are about my own raw feels or those of other humans or animals. I have allowed for the possibility that the "hard data" (i.e., those data which we can talk about with a minimum of inference) are not preanalytically but only postanalytically "given." But on just what level of psychological, introspective, phenomenological, or logical analysis we find those data which stand in the required one-one correspondence to neural events, is an open question. With W. Köhler I am inclined to think that an analysis which stops at a relatively simple configurational level (but does not proceed further to "atomize" the given) may well yield the desired items on the ψ-side of the ψ-Φ isomorphism. But phenomenal description, even of the Gestalt type, is no easy matter.

B. Unitary or Dual Language Reconstruction? In most of the crucial parts of the present essay I have taken a *unitary* language to be the ideal medium of epistemological reconstruction. By this I mean the following: Both the phenomenal terms (designating raw feel data) and the illata terms (designating unobservables) occur in the language of commonsense or of science, and they are connected by strands in the nomological net. I believe that if this sort of unitary language is constructed with care, category mistakes can be avoided. This reconstruction differs essentially from the dual language reconstruction pursued by Carnap

* For some impressive arguments against atomism see W. Köhler (183, 184, 185), Brunswik (56), Wallraff (340); and against immediacy, Lean (193), Chisholm (75), Wittgenstein (357), Rhees (278), Quinton (270), W. Sellars (315). Others like Ryle (294), Black (38) and Quine (268) have denied the possibility of a phenomenal language altogether. W. Sellars admits phenomenal concepts only as theoretical terms in a language of behavior theory.

and W. Sellars (cf. their essays 73, 315). Purely phenomenal terms are there excluded, presumably owing to their conviction that category mistakes as well as solipsism would be unavoidable if we chose a phenomenal basis of reconstruction. But with the reinstatement of realism, i.e., with the insistence on the *synthetic* character of the strands in the nomological net, solipsism is no longer a consequence, and category mistakes can be avoided if we dismiss pictorial appeals as cognitively irrelevant, and if we take care to distinguish sharply between universals and particulars, among phenomenal as well as among non-phenomenal terms.

I admit, of course, that there are certain distinct advantages in the dual language reconstruction. All evidential statements are there couched in terms of the observation language; and the observation language is conceived as *intersubjectively* meaningful right from the beginning. The connections between the observation language and the theoretical language are formulated with the help of correspondence rules. This type of reconstruction is very illuminating in the analysis of the meaning and the confirmation of scientific theories. But, as I have pointed out, it does not do full justice to statements *about* the data of direct experience, whether they are one's own or someone else's. In our unitary language the "partition" between the data and the illata is located very differently. The correspondence rules in the unitary language would ultimately be statements of ψ-Φ correlations, i.e., of the raw-feel denotations of neurophysiological terms. Since precise knowledge of these correlations is only a matter of hope for a future psychophysiology, the unitary language is largely in the "promissory note" stage. It is therefore not very illuminating if our epistemological reconstruction is to reflect the progress of knowledge in our very unfinished and ongoing scientific enterprise. For this purpose, the dual language reconstruction is much more adequate.

But if we are satisfied with relatively low probabilities for the strands in the nomological net, the unitary reconstruction might do the job too. As a sketch for a reconstruction of an ideally finished science, however, the unitary language approach is preferable. What this would amount to can at present be indicated only by some sort of "science-fiction" illustration: Suppose that we had a complete knowledge of neurophysiology and that we could order all possible human brain states (if not metrically, then at least topologically) in a phase space of *n* dimensions. Every point in this phase space would then represent a fully specific type of brain state. And, taking isomorphism for granted, a subset of these points would also represent the total set of possible mental states.

Suppose further that we could teach children the vocabulary of the language of brain states. If this requires n-tuples of numbers, then simple expressions like "17-9-6-53-12" (or even abbreviatory symbols for these) might be inculcated in the child's language. If we took care that these expressions take the place of all introspective labels for mental states, the child would immediately learn to speak about his own mental states in the language of neurophysiology. Of course, the child would not know this at first, because it would use the expression, e.g., "17-9-6-53-12" as we would "tense-impatient-apprehensive-yet hopefully-expectant." But having acquired this vocabulary, the child, when growing up and becoming a scientist, would later have no trouble in making this terminology coherent with, and part of, the conceptual system of neurophysiology, and ultimately perhaps with that of theoretical physics. Of course, I not only admit, but I would stress, that in this transformation there is a considerable change in the meaning of the original terms. But this change may be regarded essentially as an enormous enrichment, rather than as a radical shift or a "crossing of ontological barriers." In other words, introspection may be regarded as an approach to neurophysiological knowledge, although by itself it yields only extremely crude and sketchy information about cerebral processes. This sort of information may concern certain Gestalt patterns, certain qualitative and semiquantitative distinctions and gradations; but it would not, by itself, contain any indication of the cerebral connections, let alone localizations.

C. One-one Correspondence and the "Riddle of the Universe." The isomorphism of the mental and the physical consists, according to our interpretation, in a one-one correspondence of elements and relations among the phenomenal data with the elements and relations among the referents of certain neurophysiological terms. And we proposed to explain this isomorphism in the simplest way possible by the assumption of the identity of phenomenal data with the referents of (some) neurophysiological terms. The question arises whether the identity view could be held if we were, for empirical reasons, forced to abandon ψ-Φ-one-one correspondence and to replace it by a doctrine of one-many correspondence. As was pointed out previously, the physicalistic predictability of the occurrence of mental states would in principle still be unique, if one-many correspondence holds true. Comparison with an example of the identification of purely physical concepts may shed some light on this issue. Macro-temperature, as thermometrically ascertained, corresponds in one-many fashion to a multitude of micro-conditions, viz., a very large set of molecular states. Strictly speaking, this correspondence holds between one state description on the macro-level with a specifiable infinite disjunction of state descriptions pertaining to the micro-level. Since, as we have also pointed out, this correspondence is *empirically* ascertained, there is here as little reason to speak of logical identity as in the ψ-Φ case. Nevertheless, we have seen that it makes sense, and what sense it makes, to regard the relation of temperature to mean molecular kinetic energy as an example of a *theoretical* identity.

In the mind-body case, just as in the temperature case, prediction of the ensuing micro- (and ultimately even macro-) constellations on the basis of information about, respectively, the mental state, or the macro-temperature state, could not be unique under the supposition of one-many correspondence. This is obvious for the temperature example in the light of the principles of statistical mechanics. Analogously, the precise behavior subsequent to the occurrence of a specified mental state would not be predictable either. This is not too disturbing by itself. After all, even if one-one correspondence held true, the neural correlates of a mental state would form only a very insignificant part of the relevant total initial conditions. Talk of identity in the case of one-many correspondence, however, would seem unjustified, because here we are (*ex hypothesi*) *acquainted* with the phenomenal datum, and the corresponding disjunction of cerebral states could not plausibly be identified with that *individual* datum.

Even if one-one correspondence is assumed, there is an intriguing objection * against the identity view. According to the view presented in section V, there is no empirically testable difference between the identity and the parallelism doctrines. We said that the step toward the identity view is a matter of *philosophical* interpretation. But, so the objection maintains, if identity is assumed, it would be *logically* impossible to have a stream of direct experience (a "disembodied mind")

* Raised in Minnesota Center for Philosophy of Science discussions by Mr. H. Gavin Alexander.

survive bodily death and decay. It is further asserted that this would not be a logically entailed consequence of parallelism. For it could well be maintained that the one-one correspondence holds only during the life of the person, but that as drastic an event as bodily death marks the limits of this correspondence. Mental states could then occur independently of physical correlates.

Thus it would seem as if our *philosophical* identity theory implied consequences which are testably different from those of parallelism. This is quite paradoxical. My tentative reply to this argument is twofold. First, ψ-Φ identification being empirical, it could of course be mistaken. But if the identity does hold, then survival is indeed logically impossible. This is logically quite analogous to the conditional: *If* the law of the conservation of energy holds, then a *perpetuum mobile* (of the "first kind") is thereby logically excluded. But, of course, the energy law has only empirical validity and might some day be refuted by cogent empirical evidence. Second, and perhaps more important, the parallelism doctrine, as I understand it, holds that there is a ψ-Φ-one-one correspondence and that this correspondence is a matter of *universal* and irreducible law. This seems to me to exclude disembodied minds just as much as does the identity thesis. I therefore think that the identity thesis is a matter of epistemological and semantic interpretation, and does not differ in empirical consequences from a carefully formulated parallelism.

Another perplexity was formulated in Leibniz's monadology, and in different form presented by E. Du Bois-Reymond as one of his famous unsolvable "riddles of the universe." If I may put the core of the puzzle in modern form, it concerns the *irreducible* (synthetic) character of the ψ-Φ correlations. Wherever we find co-existential or correlational regularities in nature, we hope to find a unitary explanation for them, and in many cases scientific theories have provided fruitful and well-confirmed explanations of this sort. But in the case of the ψ-Φ correlation we seem to be confronted with a fundamentally different situation. There is no plausible *scientific* theory anywhere in sight which would explain just why phenomenal states are associated with brain states. Many philosophers have resigned themselves to regard the ψ-Φ correlations as "ultimate," "irreducible," "brute facts." Since any explanation presupposes explanatory premises which at least in the context of the given explanation must be accepted, and since even the introduction of higher explanatory levels usually reaches its limit after three or four "steps up," one might as well reconcile oneself to the situation, and say that "the world is what it is, and that's the end of the matter." Now, I think that it is precisely one of the advantages of the identity theory that it removes the duality of two sets of correlated events, and replaces it by the much less puzzling duality of two ways of knowing the *same* event—one direct, the other indirect.

Nevertheless, there are some "brute facts" also according to the identity theory. But they are located differently. Besides the basic physical laws and initial conditions, there are according to our view the only empirically certifiable identities of denotation of phenomenal and of physical terms. But this identity cannot be formulated in laws or lawlike sentences or formulas. The identity amounts merely to the common reference of acquaintance terms on the one hand and unique physical descriptions on the other. Any other way of phrasing the relation creates gratuitous puzzles and avoidable perplexities. For example, it is misleading to ask, "Why does a mental state 'appear' as a brain state to the physiologist?" The brain-state-as-it-appears-to-the-physiologist * is of course analyzable into phenomenal data forming part of the direct experience of the physiologist. The "brute fact" simply consists in this, that the phenomenal qualities known by acquaintance to one person are known (indirectly) by description to another person on the basis of phenomenal (evidential) data which, in the vast majority of cases, are qualitatively quite different from the data had by, or ascribed to, the first person. I see nothing paradoxical or especially puzzling in this account of the matter.

A little reflection upon the autocerebroscopic situation shows clearly that the correspondence between, e.g., musical-tones-as-directly-experienced and certain excitation patterns in the temporal lobes of one's brain as represented by visual patterns (perceived on the screen) is simply a correlation between patterns in two phenomenal fields. The conceptual neurophysiological account of the visual data in this case consists in explanatory hypotheses about cerebral processes which are *causally* responsible for the production of the image on the screen, and these are in turn causally responsible for the emergence of certain patterns in the visual field. Strictly speaking, and in the light of physical

* No matter whether the physiologist observes someone else's brain, or—autocerebroscopically—his own.

laws, there must even be a minute time lag between the moment of the occurrence of a neural event in the temporal lobe and its "representation" via the autocerebroscope in one's own visual field. The experienced patterns in the visual field are in this situation the *causal* consequences of (among other things) the auditory data. Disregarding the small time lag we could here speak of a parallelism indeed. But this is a parallelism between the data (or patterns) in different sense modalities; or, in the case of visual experience autocerebroscopically "represented" by other visual data, within one and the same modality. (May I leave it to the reader to think this through and to find out for himself that this special case of autocerebroscopy does not involve any paradoxical consequences.)

Another puzzle that may be raised is the question as to whether the proposed identity theory does not involve the undesirable consequences of *epiphenomenalism*. It should be obvious by now that our solution of the mind-body problem differs quite fundamentally from materialistic epiphenomenalism in that: (1) it is *monistic*, whereas epiphenomenalism is a form of dualistic parallelism; (2) the "physical" is interpreted as a conceptual system (or as the realities described by it), but not as the primary kind of existence, to which the mental is appended as a causally inefficacious luxury, or "shadowy" secondary kind of existence; (3) quite to the contrary, mental states experienced and/or knowable by acquaintance are interpreted as the very realities which are also denoted by a (very small) subset of physical concepts. The efficacy of pleasure, pain, emotion, deliberation, volitions, etc. is therefore quite definitely affirmed. In this respect monism shares the tenable and defensible tenets, without admitting the objectionable ones, of interactionism.

Speaking "ontologically" for the moment, the identity theory regards sentience (qualities experienced, and in human beings knowable by acquaintance) and other qualities (unexperienced and knowable only by description) the basic reality. In avoiding the unwarranted panpsychistic generalization, it steers clear of a highly dubious sort of inductive metaphysics. It shares with certain forms of idealistic metaphysics, in a very limited and (I hope) purified way, a conception of reality and combines with it the tenable component of materialism, viz., the conviction that the basic laws of the universe are "physical." This means especially, that the teleology of organic processes, the goal directedness or purposiveness of behavior are *macro*-features, and that their explanation can be given in terms of non-teleological concepts and laws which hold for the underlying micro-levels. In other words, the monistic theory here proposed does not require irreducibly teleological concepts in its explanatory premises.

In this connection there is, however, a perplexity which may give us pause. Inasmuch as we consider it a matter of empirical fact and hence of logical contingency just which physical (neurophysiological) concepts denote data of direct experience (raw feels), one may wonder whether the causal efficacy of raw feels is satisfactorily accounted for. There are countless teleological processes in organic life which, unless we be panpsychists or psychovitalists, must be regarded as occurring without the benefit of sentience. For examples, consider the extremely "ingenious" processes of reproduction, growth, adaptation, restitution, and regeneration, which occur in lower organisms as well as in many parts of human organisms. On the other hand, the causal efficacy of attention, awareness, vigilance, pleasure, pain, etc. on the human level is so striking that one is tempted, with the panpsychists, to assume some unknown-by-acquaintance qualities quite cognate with those actually experienced.

The new puzzle of epiphenomenalism would seem to come down to this: An evolutionary, physiological, and possibly physical explanation of adaptation, learning, abient or adient, goal-directed behavior can be given without any reference whatever to raw feels. The distribution of raw feels over the various possible neural states could be entirely different from what in fact it is. For example, raw feels might be associated with the peristaltic movements of the stomach or with coronary self-repair, and not with cortical processes. But, I repeat, such different distribution of raw feels or even their complete absence would still not prevent an adequate explanation of teleological behavior. Of course if we accept the *actual* distribution, i.e., the total set of ψ-Φ-correlation rules as ultimate parallel laws, and interpret these according to the identity theory, then we can quite legitimately speak of the efficacy of raw feels. This is so, because the raw-feel terms are then precisely in those loci of the nomological net where science puts (what dualistic parallelism regards as) their neural correlates. But if the biopsychological explanations offered by the theories of evolution and of learning can thus incorporate the efficacy of raw feels, those theories *presuppose*, but do not by themselves explain, the ψ-Φ correlations.

That pleasure or satisfaction reinforces certain forms of adient behavior can be formulated in the manner of the law of effect (cf. Meehl, 220). But in the ultimate neurophysiological derivation of this empirical law of behavior, the correlation of pleasure or gratification with certain cerebral states is not required. Behaviorists, especially "logical behaviorists," have taken too easy a way out here in simply *defining* the pleasurable as the behaviorally attractive and the painful as the behaviorally repellent. The "illumination" of certain physically described processes by raw feels is plainly something a radical behaviorist cannot even begin to discuss. But if the *synthetic* element in the ψ-Φ relations that we have stressed throughout is admitted, then there is something which purely physical theory does not and cannot account for. Is there then a kind of "brute fact" which our monistic theory has to accept but for which there is possibly no explanation, in the same sense as there can be (within a naturalistic empiricism) no explanation for the fact that our world is what it is in its basic laws and conditions? Possibly, however, I see a riddle here only because I have fallen victim to one of the very confusions which I am eager to eliminate from the mind-body problem. Frankly, I suspect some sort of "regression" rather than "repression" has engendered my bafflement. If so, I should be most grateful for "therapeutic" suggestions which would help in clearing up the issue. Possibly, the solution may be found in a direction which appears plausible at least for the somewhat related puzzle of the "inverted spectrum."

This ancient conundrum, we have seen, is not satisfactorily "dissolved" by Logical Behaviorism. A "captive mind" is logically conceivable, and might know by acquaintance that his sense qualia do not stand in one-one correspondence to his autocerebroscopically ascertained neural states. If physical determinism is assumed, then it is true that such knowledge would have to remain forever private and uncommunicable. But under these conditions a systematic interchange of the qualia for one person at different times and as between different persons is *logically* conceivable. It would of course *ex hypothesi* not be intersubjectively confirmable, and thus never be a possible knowledge claim of *science*. But the logical conceivability of the inverted spectrum situation demonstrates again the *empirical* character of the ψ-Φ correspondence. This empirical character is, however, (as we have also emphasized) extremely fundamental in that it is closely bound up with the basic principle of causality or of "sufficient reason." Systematic interchange of qualia for the same sort of neural states would be something for which, *ex hypothesi*, we could not state any good reasons whatever.

Furthermore, there is a grave difficulty involved in the assumption that a captive mind could even "privately" know about the interchange. Normal recollection by memory presumably involves (at least) quasi-deterministic neural processes. The captive mind could be aware of the inverted spectrum type of interchange of qualia only if we assume some peculiar breach in normal causality. If the captive mind is to *know* that today the correlation of raw feels with neural states differs from what it was yesterday, he would have to *remember* yesterday's correlations. But how could this be possible if memory depends upon modifications in the neural structures of the cortex? These considerations show clearly that under the supposition of normal physical causality the systematic interchange would remain unknowable even to the private captive mind. (Converse, but otherwise analogous, puzzles arise for the assumption of the survival of a private stream of experience beyond bodily death. How could such a private mind have knowledge about the continuance of his "physical" environment?)

All these reflections seem to me to indicate that in *our* world at least, there is nothing that is in principle inaccessible by "triangulation" on an intersubjective (sensory) basis. The *having* of raw feels is not knowledge at all, and knowledge by acquaintance does not furnish any truths which could not in principle also be confirmed indirectly by persons other than the one who verifies them directly. The ψ-Φ-identity theory as I understand it, makes explicit this "ontological" feature of our world. The criterion of scientific meaningfulness formulated in terms of intersubjective confirmability, far from being an arbitrary decree or conventional stipulation, may thus be viewed as having ontological significance—but "ontological" in the harmless sense of reflecting an inductively plausible, basic characteristic of our world.

Empirical identity, as I conceive it, is "weaker" than logical identity but "stronger" than accidental empirical identity, and like theoretical identity stronger than nomological identity in the physical sciences (just as causal necessity is weaker than logical necessity, but stronger than mere empirical regularity). If the coreference of a phenomenal term with a neurophysiological term is conceived as something more than mere extensional equivalence, if it is conceived as characteristic of the basic nature of our world (just as the basic natural laws characterize our kind of world and differentiate it from other kinds), then perhaps the inference from a neural state to its ("correlated") raw feel is at least as "necessary" (though of course not purely deductive) as is the inference from, e.g., the atomic structure of a chemical compound to its macro-physical and chemical properties.

I hope that readers sympathetic to my admittedly speculative gropings will try to formulate in logically more precise and lucid form what I have been able to adumbrate only so vaguely. Such readers should in any case keep in mind one of the ideas which seem to me indispensable for an adequate solution of the phenomenalism-realism as well as the mind-body problems: The paradigm of symbolic designation and denotation is to be seen in the relation of a token of a phenomenal term to its raw-feel referent. All non-phenomenal descriptive terms of our language, i.e., all physical terms (no matter on which level of the explanatory hierarchy) designate (or denote) entities which—within the frame of physical knowledge—are unknown by acquaintance. But if our "hypercritical" realism is accepted, we must ascribe denotata to all those physical terms which designate individuals, properties, relations, structures, fields, etc., i.e., entities which can justifiably be said to be *described* (i.e., uniquely characterized) on the basis of evidential data by Russellian descriptions on one or the other level in the hierarchy of logical types. "To exist" means simply to be the object of a true, uniquely descriptive statement. But since such descriptive knowledge (on a sensory evidential basis) by itself never enables us *deductively* to infer the acquaintance qualities of its objects, there is always a possibility for some sort of *modal* identification of a *datum* with a specifiable *descriptum*. This is the central contention of the present essay.

D. Some Remarks on the Philosophical Relevance of Open Scientific Questions in Psychophysiology. There are many problems of predominantly *scientific* character among the various mind-body puzzles. These await for their solution the further developments of biology, neurophysiology, and especially of psychophysiology.* We have touched on

* The following works and articles strike me as especially important, or at least suggestive, in these fields: Boring (40); Köhler (183, 184); Wiener (349); Hebb (145); Herrick (154); Adrian (3, 4, 5); Brain (46); Eccles (92); Ashby (9); McCulloch (214, 215); von Foerster (122, 123); Blum (39); Brillouin (49); Culbertson (80); Colby (76); Gellhorn (132); Krech (188, 189). Northrop's (240) exuberant and enthusiastic appraisal of the significance of cybernetics for the mind-body prob-

many of these issues in various parts and passages of the present essay. Speaking (again for ease of exposition only) the language of parallelism, there are, e.g., the following issues to be decided by further research concerning the specific ψ-Φ correspondences:

1. The problem of the cerebral *localization* of mental states and functions: Classical and recent experiments indicate quite specific localization for many processes. On the other hand, the findings of Lashley, Köhler, and others demonstrate a principle of mass action or of the equipotentiality of various cerebral domains.

2. The problem of the relation of phenomenal (visual, tactual, kinaesthetic, auditory, etc.) spatialities to physical space: The time-honored puzzle regarding (Lotze's) "local signs" is, as far as I know, not completely resolved. The question is by what neural mechanisms are we able to localize narrowly circumscribed events (like sensations of touch or of pain) more or less correctly on our skin or within our organism? Can we assume projection areas in the cortex which through learning processes come to interconnect afferent neural impulses in the different sensory modalities, and thus enable us to localize, e.g., visually what is first given as a tactual or pain sensation?

3. The problem of the nature of memory traces: Current fashion makes much of the reverberating circuits in neural structures. But it seems that while this explanation may do for short-range memory, it is probably not sufficient for long-range memory. Whether the lowering of neural or synaptic resistance is to be explained by "neurobiotaxis," by thickenings of the bud ends of dendrites, or by some chemical (quantum-dynamical) change in the neurons, is at present quite dubious.

4. The problem of the "specious present": The fact that the direct experience of one conscious moment embraces the events in a short stretch of finite duration, and not just an "infinitesimal" of physical time, presents a puzzle that is intriguing especially from a philosophical point of view. It is difficult, but I think not impossible to conceive of scanning mechanisms which "take note" of freshly accumulated traces, and even involve an extrapolative aspect as regards the immediate future.

5. The problem of the recollection of ordered sequences of past experiences: How can a brain process at a given time provide a correct

lem indicates at least one philosopher's response to the challenge of this new borderland discipline.

simultaneous representation of such a sequence? Philosophers are used to distinguishing a sequence of remembrances from the remembrance of a sequence of events. It seems that the latter can in certain instances occur in one moment of the specious present. Thus I seem to be aware of the sequence of themes and developments in the first movement of Beethoven's Seventh Symphony, and this awareness does not seem to require a quick internal rehearsal. It seems to be "all there at once." I also can, usually with fair reliability, recall the temporal sequence of many events in my life (various voyages, lecture engagements, first, second, third, etc. visits to Paris, and so on). Is it again some sort of "scanning" mechanism which might account for this? Driesch (87) considered it outright impossible to conceive *any* neurophysiological mechanism which would explain these phenomena, and believed that only a dualistic interactionism (involving a strictly immaterial mind or self, consonant with the rest of his vitalistic doctrines) could render justice to them. While I know of no obviously workable neural model that would do the trick, I think that Driesch, here as elsewhere, declared the defeat of naturalistic explanations prematurely. Present-day scientific findings and scientific theorizing have in so many cases shown the feasibility of physicochemical explanations of biological phenomena, so that we have good reasons to expect a successful solution of the problem of remembrance of past event-sequences.

6. The problems of "quality," "fusion," and "thresholds": I have dealt with these as best I could above (section V *E*), but there is no doubt that future research is needed in order to provide an adequate explanation for these striking phenomena.

7. The problems of "wholeness" (Gestalt), teleological functioning and purposive behavior: These also were discussed above (section IV *E*). The contributions of Gestalt theory and its doctrine of isomorphism have been largely absorbed in current psychophysiology (cf. especially Hebb, 145). Similarly significant and hopeful are the analyses of negative feedback processes as provided by cybernetics. The doctrines of "General Systems Theory," though related in spirit to cybernetics, Gestalt theory, and mathematical biophysics, are however very dubious from a logical point of view (cf. Buck, 57). We have also discussed the related issue of emergent novelty. If "absolute emergence" (Pap, 244) is a fact, then perhaps some such account as that given by Meehl and Sellars (221) may be considered seriously. I still expect that future

scientific research will demonstrate the sufficiency of $physical_2$ explanations. But if I should be wrong in that, a theory involving genuine emergence would seem to be a much more plausible alternative than dualistic interactionism. Such a theory would, however, have important philosophical implications. Inference to mental states would rest on presupposed nomological relations between $physical_2$ brain events and mental states which could be defined only in terms of the *theoretical* concepts of a $physical_1$ language. There would still be empirical identity between the referents of some (theoretical) $physical_1$ terms and the referents of phenomenal terms, but the scientific explanation of behavior would be markedly different from purely $physical_2$ explanations. Some of the philosophical puzzles of the mind-body problem might be resolved even more plausibly under this hypothesis. For example, the question regarding the "inverted spectrum" could be answered, quite straightforwardly, on the basis of normal inductive or analogical inference. Directly given qualia, represented by (theoretical) $physical_1$ terms in our scientific account would then be functionally related to those brain processes which are described in $physical_2$ (theoretical) terms. The principle of sufficient reason would then tell us that to assume any deviation from the highly confirmed functional relationships between mental states and $physical_2$ brain states would be just as arbitrary as, e.g., the assumption that some electric currents are associated with magnetic fields of an entirely different structure than are others (despite the complete similarity of the electric currents in every other respect). As I have indicated before, the validity of the emergentist theory falls in any case under the jurisdiction of future empirical research.

8. The problem of a neurophysiological account of *selfhood*: This important though controversial notion describes a form of organization or integration of experiences and dispositions which on the neural side corresponds first to the relatively stable structure of the brain and the other parts of the nervous system, as well as to certain unified forms of functioning. To what extent the psychoanalytic concepts of the ego, superego, and id may be "identified" with such structures and functions is still very unclear. Very likely, the psychological notions will appear only as first crude approximations, once the detailed neurophysiological facts are better known.

9. The problems of neurophysiological theories which will account for the *unconscious* processes assumed by various "depth psychologies,"

especially psychoanalysis: One of the philosophically intriguing questions here is whether we can explicate such psychoanalytic concepts as "repressed wishes", "unconscious anxiety", "Oedipus complex", etc. as *dispositions*, or whether unconscious *events* also need to be assumed. Even outside the sphere of Freudian preoccupations, there are for instance the often reported cases of "waking up with the solution of a mathematical problem." One wonders whether the brain did some "work" during sleep, and if so, whether "unconscious thoughts" might not be part of a first-level explanation of this sort of phenomenon. I am inclined to think that *both* dispositions *and* events are required, and that the future development of science may well produce more reliable neurophysiological explanations than the currently suggested (and suggestive) brain models (cf. Colby, 76).

10. Much more problematic than all the questions so far discussed in this section are the implications of the alleged findings of *psychical research*. Having been educated in the exercise of the scientific method, I would in the first place insist on further experimental scrutiny of those findings. But if we take seriously the impressive statistical evidence in favor of telepathy, clairvoyance, and precognition, then there arises the extremely difficult problem of how to account for these facts by means of a scientific theory. I know of no attempt that gives even a plausible suggestion for such a theory. All hypotheses that have been proposed so far are so utterly fantastic as to be scientifically fruitless for the present. But logical analyses (e.g., C. D. Broad, 52; M. Scriven, 304) which make explicit in which respects the facts (*if* they are facts!) of psychical research are incompatible with some of the guiding principles of ("Victorian"!) science are helpful and suggestive. It is difficult to know whether we stand before a scientific revolution more incisive than any other previous revampings of the frame of science, or whether the changes which may have to be made will only amount to minor emendations.

Concluding Remark. An essential part of the justification of the philosophical monism proposed in this essay depends upon empirical, scientific assumptions. Only the future development of psychophysiology will decide whether these assumptions are tenable. Since I am not a laboratory scientist (though I did some laboratory work in physics and chemistry in my early years), I cannot responsibly construct psychophysiological hypotheses. Nor did I intend to close the doors to alternative philosophical views of the relations of the mental to the physical. What I did try to show, however, is that monism is

(1) still very plausible on scientific grounds,
(2) philosophically defensible in that it involves no insurmountable logical or epistemological difficulties and paradoxes.

I realize fully that I could deal only with some of the perplexities which have vexed philosophers or psychologists throughout the ages, and especially in recent decades. Just where the philosophical shoe pinches one, just which problems strike one as important—that depends, of course, on a great many more or less accidental personal, educational, or cultural factors. Despite my valiant efforts to deal with what strike *me* as important and baffling questions, I may of course not even have touched on other facets which some of my readers might consider as *the* essential problems of mind and body. May others come and deal with them!

Bibliography

In the following rather ample bibliography I have tried to assemble not only references to those materials actually discussed or quoted in my essay, but also a great deal of what seemed to me of systematic significance for future philosophical work in the area. With the appalling volume of philosophical writings in recent decades, many a valuable book or article becomes all too soon forgotten, and many go entirely unnoticed. Scholars or students who wish to tackle the "world knot" may find most of these books or articles stimulating, and many of them illuminating.

As regards my earlier publications on the mind-body problem, I now regard my presentation (103) of 1934 as partly confused. The later rather compact presentation (112) of 1950 presents on the whole an adequate preview and summary of my present outlook (though I am no longer satisfied with some of the illustrative analogies used there). A fuller discussion of my identity theory in relation to Carnap's present (largely unpublished) version of physicalism and to the issues of the empiricist criterion of factual meaningfulness is contained in my essay (116) in the Carnap volume of P. A. Schilpp's Library of Living Philosophers.

1. Adams, D. K. "Learning and explanation," *Learning Theory, Personality Theory, and Clinical Research: The Kentucky Symposium*, pp. 66–80. New York: Wiley; London: Chapman & Hall, Ltd., 1954.
2. Adams, G. P., J. Loewenberg, and S. C. Pepper (eds.). *The Nature of Mind*, University of California Publications in Philosophy, Vol. 19, 1936 (especially articles by P. Marhenke and S. C. Pepper).
3. Adrian, E. D. *The Mechanism of Nervous Action*. Philadelphia: Univ. of Pennsylvania Press, 1932.
4. Adrian, E. D. *The Physical Background of Perception*. Oxford: The Clarendon Press, 1947.
5. Adrian, E. D., F. Bremer, H. H. Jasper (consulting eds.), and J. F. Delafresnaye (ed. for the Council). *Brain Mechanisms and Consciousness: a Symposium—Council for International Organizations of Medical Sciences*. Springfield (Ill.): Charles C. Thomas, 1954.
6. Aldrich, V. C. "Messrs. Schlick and Ayer on Immortality," *Philosophical Review*, 47:209–213 (1938). Reprinted in H. Feigl and W. Sellars, *Readings in Philosophical Analysis*. New York: Appleton-Century-Crofts, 1949.
7. Aldrich, V. C. "What Appears?" *Philosophical Review*, 63:232–240 (1954).
8. Alexander, P. "Complementary Descriptions," *Mind*, 65:145–65 (1956).
9. Ashby, W. R. *Design for a Brain*. New York: Wiley, 1952.
10. Austin, J. L. "Other Minds," *Aristotelian Society Supplementary Volume*, 20: 148–187 (1946). Reprinted in A. G. N. Flew, *Logic and Language* (2nd series) 123–58. Oxford: Blackwell; New York: Philosophical Library, 1953.
11. Avenarius, R. *Der Menschliche Weltbegrift*. Leipzig: Reisland, 1912.
12. Ayer, A. J. *Language, Truth and Logic*. New York: Oxford Univ. Press, 1936; 2nd ed., London: Gollanz, 1946.

13. Ayer, A. J. *The Foundations of Empirical Knowledge*. New York: Macmillan, 1940.
14. Ayer, A. J. "The Physical Basis of Mind: A Philosophers' Symposium II," in Peter Laslett (ed.), *The Physical Basis of Mind*, pp. 70–74. New York: Macmillan, 1950.
15. Ayer, A. J. "One's Knowledge of Other Minds," *Theoria*, 19:1–20, Part 1–2 (1953).
16. Ayer, A. J. "Can There be a Private Language?" A Symposium, *Aristotelian Society Supplementary Volume*, 28:63–76 (1954).
17. Ayer, A. J. "Phenomenalism," in A. J. Ayer, *Philosophical Essays*. New York: St. Martin's Press, 1954.
18. Ayer, A. J. *The Problem of Knowledge*. New York: St. Martin's Press, 1956.
19. Bakan, D. "A Reconsideration of the Problem of Introspection," *Psychological Bulletin*, 51:106–118, March 1954.
20. Bar-Hillel, Y. "Indexical Expressions," *Mind*, 63:359–379 (1954).
21. Beck, L. W. "The Psychophysical as a Pseudo-Problem," *Journal of Philosophy*, 37:561–571 (1940).
22. Beck, L. W. "The Principle of Parsimony in Empirical Science," *Journal of Philosophy*, 40:617–633 (1943).
23. Beck, L. W. "Secondary Quality," *Journal of Philosophy*, 43:599–610 (1946).
24. Beck, L. W. "Constructions and Inferred Entities," *Philosophy of Science*, 17: 74–86 (1950). Reprinted in H. Feigl and M. Brodbeck (eds.), *Readings in the Philosophy of Science*, pp. 368–381. New York: Appleton-Century-Crofts, 1953.
25. Berenda, C. W. "On Emergence and Prediction," *Journal of Philosophy*, 50: 269–274 (1953).
26. Bergmann, G. "On Physicalistic Models of Non-physical Terms," *Philosophy of Science*, 7:151–158 (1940).
27. Bergmann, G. "On Some Methodological Problems of Psychology," *Philosophy of Science*, 7:205–219 (1940). Reprinted in H. Feigl and M. Brodbeck (eds.), *Readings in the Philosophy of Science*, pp. 627–636. New York: Appleton-Century-Crofts, 1953.
28. Bergmann, G. "Holism, Historicism and Emergence," *Philosophy of Science*, 11:209–221 (1944).
29. Bergmann, G. "The Logic of Psychological Concepts," *Philosophy of Science*, 18:93–110 (1951).
30. Bergmann, G. "The Problem of Relations in Classical Psychology," *Philosophical Quarterly*, 2:140–152 (1952).
31. Bergmann, G. "Theoretical Psychology," *Annual Review of Psychology*, 4:435–458 (1953).
32. Bergmann, G. "Sense and Nonsense in Operationism," *Science Monthly*, 79: 210–214 (1954).
33. Bergmann, G. "Intentionality," *Archivio di Filosofia*, 6:177–216 (1955).
34. Bergmann, G. *Philosophy of Science*. Madison: Univ. of Wisconsin Press, 1957.
35. Berlin, I. "Empirical Propositions and Hypothetical Statements," *Mind*, 59:289–312 (1950).
36. Bichowsky, F. R. "Factors Common to the Mind and to the External World," *Journal of Philosophy*, 37:18 (1940).
37. Black, M. "Linguistic Method in Philosophy," *Philosophical and Phenomenological Research*, 8:4 (1948). Reprinted in M. Black, *Language and Philosophy*. Ithaca: Cornell Univ. Press, 1949.
38. Black, M. "Symposium: Phenomenalism." *Science, Language, and Human Rights*. American Philosophical Association, Vol. 1, Philadelphia: Univ. of Pennsylvania Press, 1952.

39. Blum, H. F. *Time's Arrow and Evolution*. Princeton: Princeton Univ. Press, 1955.
40. Boring, E. G. *The Physical Dimensions of Consciousness*. New York, London: The Century Co., 1933.
41. Boring, E. G. "Psychophysiological Systems and Isomorphic Relations," *Psychological Review*, 43:565–587 (1936).
42. Boring, E. G. "A Psychological Function is the Relation of Successive Differentiations of Events in the Organism," *Psychological Review*, 44:445–461 (1937).
43. Boring, E. G., *et al.* Symposium on Operationism. *Psychological Review*, 52: 241–294 (1945).
44. Boring, E. G. "Mind and Mechanism," *American Journal of Psychology*, 59: 173–192 (1946).
45. Boring, E. G. "A History of Introspection," *Psychological Bulletin*, 50:169–189 (1953).
46. Brain, W. R. *The Contribution of Medicine to Our Idea of Mind*. Cambridge: Cambridge Univ. Press, 1952.
47. Braithwaite, M. "Causal Laws in Psychology," A Symposium, *Aristotelian Society Supplementary Volume*, 23:45–60 (1949).
48. Braithwaite, R. B. *Scientific Explanation*. Cambridge: Cambridge Univ. Press, 1953.
49. Brillouin, L. *Science and Information Theory*. New York: Academic Press, 1956.
50. Broad, C. D. *Perception, Physics, and Reality*. Cambridge: Cambridge Univ. Press, 1914.
51. Broad, C. D. *The Mind and Its Place in Nature*. London: Routledge & Kegan Paul, 1925.
52. Broad, C. D. "The Relevance of Psychical Research to Philosophy," *Philosophy*, 24:291–309 (1949). Reprinted in C. D. Broad, *Religion, Philosophy, and Psychical Research*. New York: Harcourt, Brace, 1953.
53. Brodbeck, M. "On the Philosophy of the Social Sciences," *Philosophy of Science*, 21:140–156 (1954).
54. Brunswik, E. "The Conceptual Focus of Some Psychological Symptoms," in P. L. Harriman (ed.), *Twentieth Century Psychology*. New York: Philosophical Library, 1946.
55. Brunswik, E. "Points of View," in *Encyclopedia of Psychology* (P. L. Harriman, ed.). New York: Philosophical Library, 1946.
56. Brunswik, E. "The Conceptual Framework of Psychology," *International Encyclopedia of Unified Science*, Vol. 10. Chicago: Univ. of Chicago Press, 1952.
57. Buck, R. C. "On the Logic of General Behavior Systems Theory," in H. Feigl and M. Scriven (eds.). *Minnesota Studies in the Philosophy of Science*, Vol. I, *The Foundations of Science and the Concepts of Psychology and Psychoanalysis*, pp. 223–238. Minneapolis: Univ. of Minnesota Press, 1956.
58. Burks, A. W. "Icon, Index and Symbol," *Philosophical and Phenomenological Research*, 9:673–689 (1949).
59. Burks, A. W. "The Logic of Causal Propositions," *Mind*, 60:363–382 (1951).
60. Carnap, R. *Der Logische Aufbau der Welt*. Berlin: Weltkreis, 1928.
61. Carnap, R. *Scheinproblem in der Philosophie*. Berlin: Weltkreis, 1928.
62. Carnap, R. "Psychologie in Physikalischer Sprache," *Erkenntnis*, 3:107–142 (1933).
63. Carnap, R. "Les Concepts Psychologiques et les Concepts Physiques, sont-ils Foncièrement Différents?" *Revue de Synthèse*, 10:43–53 (1935).
64. Carnap, R. "Testability and Meaning," *Philosophy of Science*, 3:420–468 (1936); 4:1–40 (1937). Also reprinted by Graduate Philosophy Club, Yale University, New Haven, Conn., 1950. Also reprinted in H. Feigl and M. Brodbeck (eds.), *Readings in the Philosophy of Science*, pp. 47–92. New York: Appleton-Century-Crofts, 1953.

65. Carnap, R. *The Logical Syntax of Language*. New York: Harcourt, Brace, 1937.
66. Carnap, R. *The Unity of Science*. London: Kegan Paul, 1938.
67. Carnap, R. "Logical Foundations of the Unity of Science," *International Encyclopedia of Unified Science*, Vol. I, No. 1 (R. Carnap and C. W. Morris, eds.). Chicago: Univ. of Chicago Press, 1938, 42–62. Reprinted in H. Feigl and W. Sellars (eds.), *Readings in Philosophical Analysis*, pp. 408–423. New York: Appleton-Century-Crofts, Inc., 1949.
68. Carnap, R. "Foundations of Logic and Mathematics," *International Encyclopedia of Unified Science*, Vol. I, No. 3. Chicago: Univ. of Chicago Press, 1939. A part ("The Interpretation of Physics") is reprinted in H. Feigl and M. Brodbeck (eds.), *Readings in the Philosophy of Science*, pp. 309–318. New York: Appleton-Century-Crofts, 1953.
69. Carnap, R. *Introduction to Semantics*. Cambridge (Mass.): Harvard Univ. Press, 1942.
70. Carnap, R. "Empiricism, Semantics, and Ontology," *Revue Internationale de Philosophie*, 4:20–40 (1950). Reprinted in P. P. Wiener (ed.), *Readings in the Philosophy of Science*, pp. 509–521. New York: Scribner, 1953. Also reprinted in L. Linsky (ed.), *Semantics and the Philosophy of Language*, pp. 208–230. Urbana (Ill.): Univ. of Illinois Press, 1952.
71. Carnap, R. "Meaning and Synonymy in Natural Languages," *Philosophical Studies*, 6:33–47 (1955).
72. Carnap, R. *Meaning and Necessity*, 2nd edition. Chicago: Univ. of Chicago Press, 1956.
73. Carnap, R. "The Methodological Character of Theoretical Concepts," in *Minnesota Studies in the Philosophy of Science*, Vol. I, pp. 38–76. Minneapolis: Univ. of Minnesota Press, 1956.
74. Castell, A. "The Critical and the Mechanical," *The Philosophical Review*, 60:1 (1951).
75. Chisholm, R. M. "Verification and Perception," *Revue Internationale de Philosophie*, 17:1–17 (1951).
76. Colby, K. M. *Energy and Structure in Psychoanalysis*. New York: Ronald Press Co., 1955.
77. Copi, I. M. "A Note on Representation in Art," *Journal of Philosophy*, 52:346–349 (1955).
78. Cory, D. "Are Sense-Data 'in' the Brain?" *Journal of Philosophy*, 65:533–548 (1948).
79. Cronbach, L. J., and P. M. Meehl. "Construct Validity in Psychological Tests," *Psychological Bulletin*, 52:281–302 (1955). Reprinted in H. Feigl and M. Scriven (eds.), *Minnesota Studies in the Philosophy of Science*, Vol. I, pp. 174–204. Minneapolis: Univ. of Minnesota Press, 1956.
80. Culbertson, J. T. *Consciousness and Behavior*. Dubuque (Iowa): W. C. Brown Co., 1950.
81. Dennes, W. R. "Mind and Meaning," in G. P. Adams, J. Loewenberg, and S. C. Pepper (eds.), *The Nature of Mind*, pp. 1–30. Berkeley: Univ. of California Press, 1936.
82. Deutsch, K. W. "Mechanism, Teleology, and Mind: The Theory of Communications and Some Problems in Philosophy and Social Science," *Philosophical and Phenomenological Research*, 12:185–223 (1951).
83. Dewey, J. *Experience and Nature*. LaSalle (Ill.): Open Court Pub. Co., 1926.
84. Dewey, J. "Conduct and Experience," in C. Murchison (ed.), *Psychologies of 1930*. Worcester (Mass.): Clark Univ. Press, 1930.

85. Drake, D. *Mind and its Place in Nature.* New York: Macmillan, 1925.
86. Drake, D. *Invitation to Philosophy.* Chicago: Houghton Mifflin Co., 1933.
87. Driesch, H. *Mind and Body.* London: Methuen, 1927.
88. Driesch, H. *Philosophische Gegenwartsfragen.* Leipzig: Verlag Emmanuel Reinicke, 1933.
89. Ducasse, C. J. "On the Attributes of Material Things," *Journal of Philosophy,* 31:57–72 (1934).
90. Ducasse, C. J. *Nature, Mind, and Death.* LaSalle (Ill): Open Court Pub. Co., 1951.
91. Ducasse, C. J. "Demos on 'Nature, Mind and Death,'" *The Review of Metaphysics,* 7:290–298 (1953).
92. Eccles, J. C. *The Neurophysiological Basis of Mind.* Oxford: The Clarendon Press, 1953.
93. Eddington, A. S. *The Nature of the Physical World.* New York: Macmillan, 1928.
94. Eddington, A. S. *Science and the Unseen World.* New York: Macmillan, 1929.
95. Elsasser, W. M. "Quantum Mechanics, Amplifying Processes, and Living Matter," *Philosophy of Science,* 18:300–326 (1951).
96. Elsasser, W. M. "A Reformulation of Bergson's Theory of Memory," *Philosophy of Science,* 20:7–21 (1953).
97. Elsasser, W. M. *The Physical Foundation of Biology.* New York: Pergamon Press, 1958.
98. Epstein, J. "Professor Ayer on Sense-Data," *Journal of Philosophy,* 53:401–415 (1956).
99. Farrell, B. A. "Causal Laws in Psychology," A Symposium, *Aristotelian Society Supplementary Volume,* 23:30–44 (1949).
100. Farrell, B. A. "Critical Notice on 'The Concept of Mind' by Gilbert Ryle," *British Journal of Psychology,* 40:159–164 (1950).
101. Farrell, B. A. "Experience," *Mind,* 59:170–198 (1950).
102. Farrell, B. A. "Intentionality and the Theory of Signs," *Philosophical and Phenomenological Research,* 15:500–511 (1955).
103. Feigl, H. "Logical Analysis of the Psychophysical Problem," *Philosophy of Science,* 1:420–445 (1934).
104. Feigl, H. "Moritz Schlick," *Erkenntnis,* 7:393–419 (1938).
105. Feigl, H. "Logical Empiricism," in D. D. Runes (ed.), *Twentieth Century Philosophy,* New York: Philosophical Library, 1943, pp. 371–416. Reprinted in H. Feigl and W. Sellars, *Readings in Philosophical Analysis.* New York: Appleton-Century-Crofts, 1949. Reprinted (in part) in M. Mandelbaum, *et al., Philosophic Problems,* pp. 3–26. New York: Macmillan, 1957.
106. Feigl, H. "Operationism and Scientific Method," *Psychological Review,* 52:250–259 (1945). Reprinted in H. Feigl and W. Sellars (eds.), *Readings in Philosophical Analysis,* pp. 498–509. New York: Appleton-Century-Crofts, 1949.
107. Feigl, H. "Naturalism and Humanism," *American Quarterly,* 1:135–149 (1949). Reprinted in H. Feigl and M. Brodbeck (eds.), *Readings in the Philosophy of Science,* pp. 8–18. New York: Appleton-Century-Crofts, 1953.
108. Feigl, H. "Some Remarks on the Meaning of Scientific Explanation," in H. Feigl and W. Sellars (eds.), *Readings in Philosophical Analysis,* pp. 510–514. New York: Appleton-Century-Crofts, 1949.
109. Feigl, H. "De Principiis Non Disputandum . . .? On the Meaning and the Limits of Justification," in M. Black (ed.), *Philosophical Analysis,* pp. 119–156. Ithaca (N.Y.): Cornell University Press, 1950.
110. Feigl, H. "Existential Hypotheses: Realistic Versus Phenomenalistic Interpretations," *Philosophy of Science,* 17:35–62 (1950).

111. Feigl, H. "Logical Reconstruction, Realism and Pure Semiotic," *Philosophy of Science*, 17:186–195 (1950).
112. Feigl, H. "The Mind-Body Problem in the Development of Logical Empiricism," *Revue Internationale de Philosophie*, 4:64–83 (1950). Reprinted in H. Feigl and M. Brodbeck (eds.), *Readings in the Philosophy of Science*, pp. 612–626. New York: Appleton-Century-Crofts, 1953.
113. Feigl, H. "Principles and Problems of Theory Construction in Psychology," in W. Dennis (ed.), *Current Trends of Psychological Theory*, pp. 174–213. Pittsburgh: Univ. of Pittsburgh Press, 1951.
114. Feigl, H. "Scientific Method Without Metaphysical Presuppositions," *Philosophical Studies*, 5:17–19 (1954). Reprinted with minor alterations in H. Feigl and M. Scriven (eds.), *Minnesota Studies in the Philosophy of Science*, Vol. I, pp. 22–37. Minneapolis: Univ. of Minnesota Press, 1956.
115. Feigl, H. "Functionalism, Psychological Theory, and the Uniting Sciences: Some Discussion Remarks," *Psychological Review*, 62:232–235 (1955).
116. Feigl, H. "Physicalism, Unity of Science and the Foundations of Psychology," in P. A. Schilpp (ed.), *The Philosophy of Rudolf Carnap*. LaSalle (Ill.): Open Court Pub. Co., 1963.
117. Feigl, H., and M. Brodbeck (eds.). *Readings in the Philosophy of Science*. New York: Appleton-Century-Crofts, 1953.
118. Feigl, H., and W. Sellars (eds.). *Readings in Philosophical Analysis*. New York: Appleton-Century-Crofts, 1949.
119. Feyerabend, P. K. "Carnap's Theorie der Interpretation theoretischer Systeme," *Theoria*, 21:55–62 (1955).
120. Feyerabend, P. K. "A Note on the Paradox of Analysis," *Philosophical Studies*, 7:92–96 (1956).
121. Findlay, J. N. "Is There Knowledge by Acquaintance?" A Symposium, *Aristotelian Society Supplementary Volume*, 23:111–128 (1949).
122. Förster, H. *Das Gedächtnis*. Vienna: Franz Deuticke, 1948.
123. Förster, H. (ed.). *Conferences on Cybernetics*, Vols. 6–9. New York: Josiah Macy, Jr. Foundation, 1949–1952.
124. Frank, L. K., *et al*. *Teleological Mechanisms*. New York: The New York Academy of Sciences, 1948.
125. Frank, P. *Modern Science and its Philosophy*. Cambridge (Mass.): Harvard Univ. Press, 1949.
126. Frank, P. G. *Philosophy of Science*. New York: Prentice-Hall, 1957.
127. Frenkel-Brunswik, E. "Psychoanalysis and the Unity of Science," *Proceedings of the American Academy of Arts and Sciences*, 80:271–350 (1954).
128. Freytag, W. *Der Realismus und das Transzendenzproblem*. Halle: Niemeyer, 1902.
129. Fritz, C. A., Jr. "Sense Perception and Material Objects," *Philosophical and Phenomenological Research*, 16:303–316 (1956).
130. Gätschenberger, R. *Symbola*. Karlsruhe: G. Braun, 1920.
131. Gätschenberger, R. *Zeichen, die Fundamente des Wissens*. Stuttgart: Frommann, 1932.
132. Gellhorn, E. *Physiological Foundations of Neurology and Psychiatry*. Minneapolis: Univ. of Minnesota Press, 1953.
133. Ginsberg, A. "Hypothetical Constructs and Intervening Variables," *Psychological Review*, 61:119–131 (1954).
134. Ginsberg, A. "Operational Definitions and Theories," *Journal of General Psychology*, 52:223–245 (1955).
135. Goodman, N. *The Structure of Appearance*. Cambridge: Harvard Univ. Press, 1951.

136. Goodman, N. "Sense and Certainty," *Philosophical Review*, 61:160–167 (1952).
137. Goodman, N. "The Revision of Philosophy," in S. Hook (ed.), *American Philosophers at Work*, pp. 75–92. New York: Criterion Books, 1956.
138. Grossmann, R. S. "Zur Logischen Analyse des Neobehaviorismus," *Psychologische Rundschau*, Band VI/4:246–260 (1955).
139. Grünbaum, A. "Operationism and Relativity," *Scientific Monthly*, 79:228–231 (1954).
140. Guthrie, E. R. "Purpose and Mechanism in Psychology," *Journal of Philosophy*, 21:25 (1924).
141. Hampshire, S. "The Analogy of Feeling," *Mind*, 61:1–12 (1952).
142. Hart, H. L. A. "Is There Knowledge by Acquaintance?" A Symposium, *Aristotelian Society Supplementary Volume*, 23:69–90 (1949).
143. Hathaway, S. R. "Clinical Intuition and Inferential Accuracy," *Journal of Personality*, 24:223–250 (1956).
144. Hayek, F. A. "The Facts of the Social Sciences," *Ethics*, 54:1 (1943).
145. Hebb, D. O. *The Organization of Behavior, A Neuropsychological Theory*. New York: Wiley, 1949.
146. Hempel, C. G. "The Logical Analysis of Psychology," *Revue de Synthèse*, 10: 27–42 (1935). Reprinted in H. Feigl and W. Sellars (eds.), *Readings in Philosophical Analysis*, pp. 373–384. New York: Appleton-Century-Crofts, 1949.
147. Hempel, C. G. "The Concept of Cognitive Significance: A Reconsideration," *Proceedings of the American Academy of Arts and Sciences*, 80:61–77 (1951).
148. Hempel, C. G. "Fundamentals of Concept Formation in the Empirical Sciences," *International Encyclopedia of Unified Science*, Vol. II, No. 7. Chicago: Univ. of Chicago Press, 1952.
149. Hempel, C. G. "Problems and Changes in the Empiricist Criterion of Meaning," *Revue Internationale de Philosophie*, 4:41–63 (1950). Reprinted in L. Linsky (ed.), *Semantics and the Philosophy of Language*. Urbana: Univ. of Illinois Press, 1952.
150. Hempel, C. G. "Reflections on Nelson Goodman's *The Structure of Appearance*," *Philosophical Review*, 62:108–116 (1953).
151. Hempel, C. G. "A Logical Appraisal of Operationism," *Scientific Monthly*, 79:215–220 (1954).
152. Hempel, C. G., and P. Oppenheim. "The Logic of Explanation," *Philosophy of Science*, 15:135–175 (1948). Reprinted in H. Feigl and M. Brodbeck (eds.), *Readings in the Philosophy of Science*, pp. 319–352. New York: Appleton-Century-Crofts, 1953.
153. Henle, P. "The Status of Emergence," *Journal of Philosophy*, 39:486–493 (1942).
154. Herrick, C. J. *The Evolution of Human Nature*. Austin: Univ. of Texas Press, 1956.
155. Hervey, H. "The Private Language Problem," *Philosophical Quarterly*, 7:63–79 (1957).
156. Heymans, G. *Einführung in die Metaphysik*. Leipzig: Barth, 1921.
157. Hobart, R. E. "Free Will as Involving Determinism and Inconceivable Without It," *Mind*, 43:1–26 (1934).
158. Hofstadter, A. "Professor Ryle's Category-Mistake," *Journal of Philosophy*, 48: 257–270 (1951).
159. Holt, E. B. *The Concept of Consciousness*. London: G. Allen and Co., Ltd., 1914.
160. Hospers, J. *An Introduction to Philosophical Analysis*. New York: Prentice-Hall, 1953.
161. Hughes, G. E. "Is There Knowledge by Acquaintance?" A Symposium, *Aristotelian Society Supplementary Volume*, 23:91–110 (1949).

162. Hull, C. L. "Mind, Mechanism, and Adaptive Behavior," *Psychological Review*, 44:1–32 (1937).
163. Jacobs, N. "Physicalism and Sensation Sentences," *Journal of Philosophy*, 34:22 (1937).
164. Jacoby, G. *Allgemeine Ontologie der Wirklichkeit*, Vol. 2. Halle: Niemeyer, 1955.
165. James, W. *Principles of Psychology*. New York: H. Holt and Co., 1890.
166. Juhos, B. *Die Erkenntnis und ihre Leistung*. Vienna: Springer-Verlag, 1950.
167. Kaila, E. "Beiträge zu einer Synthetischen Philosophie," *Annales Universitatis Aboensis*, 4:9–208 (1928).
168. Kaila, E. "Physikalismus und Phänomenalismus," *Theoria*, 8:85–125 (1942).
169. Kaila, E. "Terminalkansalität als die Grundlage eines Unitarischen Naturbegriffs," *Acts Philosophica Fennica*, 10:7–122 (1956).
170. Kaplan, A. "Definition and Specification of Meaning," *Journal of Philosophy*, 63:281–288 (1946).
171. Kaplan, A. and H. F. Schott. "A Calculus for Empirical Classes," *Methodos*, 3:165–190 (1951).
172. Kapp, R. O. *Science vs. Materialism*. London: Methuen, 1940.
173. Kapp, R. O. *Mind, Life, and Body*. London: Constable, 1951.
174. Kapp, R. O. *Facts and Faith: The Dual Nature of Reality*. New York: Oxford Univ. Press, 1955.
175. Kaufmann, F. *Methodology of the Social Sciences*. New York: Oxford Univ. Press, 1944.
176. Keller, F. S. *The Definition of Psychology*. New York: Appleton-Century-Crofts, 1937.
177. Kemeny, J. G., and P. Oppenheim. "On Reduction," *Philosophical Studies*, 7:6–19 (1956).
178. King, H. R. "Professor Ryle and the Concept of Mind," *Journal of Philosophy*, 48:9 (1951).
179. Kneale, W. "Induction, Explanation, and Transcendent Hypotheses," in W. Kneale, *Probability and Induction*, pp. 92–110. Oxford: The Clarendon Press, 1949. Reprinted in H. Feigl and M. Brodbeck (eds.), *Readings in the Philosophy of Science*, pp. 353–367. New York: Appleton-Century-Crofts, 1953.
180. Kneale, W. *Probability and Induction*. Oxford: The Clarendon Press, 1949.
181. Köhler, W. "Bemerkungen zum Leib-Seele-Problem," *Deutsche Medizinische Wochenschrift*, 50:1269–1270 (1924).
182. Köhler, W. "Ein altes Scheinproblem," *Die Naturwissenschaften*, 17:395–401 (1929).
183. Köhler, W. *The Place of Values in a World of Facts*. New York: Liveright, 1938.
184. Köhler, W. *Dynamics in Psychology*. New York: Liveright, 1940.
185. Köhler, W. *Gestalt Psychology*. New York: Liveright, 1947.
186. Köhler, W. "Direction of Processes in Living Systems," in P. G. Frank (ed.), *The Validation of Scientific Theories*, pp. 143–150. Boston: Beacon Press, 1954.
187. Kraft, V. *The Vienna Circle* (translated by A. Pap). New York: Philosophical Library, 1953.
188. Krech, D. "Dynamic Systems, Psychological Fields, and Hypothetical Constructs," *Psychological Review*, 57:283–290 (1950).
189. Krech, D. "Dynamic Systems as Open Neurological Systems," *Psychological Review*, 57:345–361 (1950).
190. Krikorian, Y. H. (ed.). *Naturalism and the Human Mind*. New York: Columbia Univ. Press, 1944.

191. Külpe, O. *Die Realisierung.* Leipzig: S. Hirzel, 1912.
192. Lashley, K. S. "Behaviorism and Consciousness," *Psychological Review,* 30:346 (1923).
193. Lean, Martin. *Sense Perception and Matter.* London: Routledge and Kegan Paul, 1953.
194. Lecomte Du Noüy, P. *Human Destiny.* New York: Longmans, Green and Co., 1947.
195. Lewis, C. I. *Mind and the World Order.* New York: Scribner, 1929.
196. Lewis, C. I. "Some Logical Considerations Concerning the Mental," *Journal of Philosophy,* 38 (1941). Reprinted in H. Feigl and W. Sellars (eds.), *Readings in Philosophical Analysis,* pp. 385–392. New York: Appleton-Century-Crofts, 1949.
197. Lewis, C. I. *An Analysis of Knowledge and Valuation.* LaSalle (Ill.): Open Court Pub. Co., 1946.
198. Lewis, C. I. "The Given Element in Empirical Knowledge," *Philosophical Review,* 61:168–175 (1952).
199. Lewy, C. "Is the Notion of Disembodied Existence Self-Contradictory?" *Proceedings of the Aristotelian Society,* 43:59–78 (1942–1943).
200. Lindzey, G. "Hypothetical Constructs, Conventional Constructs, and Use of Physiological Data in Psychological Theory," *Psychiatry,* 16:27–33 (1953).
201. Littman, R. A. "Mr. Ryle on 'Thinking,'" *Acta Psychologica,* 10:381–384 (1954).
202. London, I. D. "Free Will as a Function of Divergence," *Psychological Review,* 55:41–47 (1948).
203. London, I. D. "Quantum Biology and Psychology," *Journal of General Psychology,* 46:123–149 (1952).
204. Lovejoy, A. O. *The Revolt Against Dualism.* New York: Norton & Co., 1930.
205. Mace, C. A. "Causal Laws in Psychology," A Symposium, *Aristotelian Society Supplementary Volume,* 23:61–68 (1949).
206. Mach, E. *The Analysis of Sensations.* LaSalle (Ill.): Open Court Pub. Co., 1914.
207. Madden, E. N. "Discussion—Science, Philosophy and Gestalt Theory," *Philosophy of Science,* 20:329–331 (1953).
208. Margenau, H. *The Nature of Physical Reality.* New York: McGraw-Hill Book Co., 1950.
209. Margenau, H. "The Exclusion Principle and Its Philosophical Importance," *Philosophy of Science,* 11:187–208 (1944).
210. Marhenke, P. "The Constituents of Mind," in G. P. Adams, J. Loewenberg, and S. C. Pepper (eds.), *The Nature of Mind,* pp. 171–208. Berkeley: Univ. of California Press, 1936.
211. Marx, M. H. (ed.). *Psychological Theory.* New York: Macmillan, 1951.
212. Maze, J. R. "Do Intervening Variables Intervene?" *Psychological Review,* 61: 226–234 (1954).
213. MacCorquodale, K., and P. E. Meehl. "On a Distinction between Hypothetical Constructs and Intervening Variables," *Psychological Review,* 55:95–107 (1948).
214. McCulloch, W. S. "Brain and Behavior," in *Current Trends in Psychological Theory,* pp. 165–178. Pittsburgh: Univ. of Pittsburgh Press, 1951.
215. McCulloch, W. S. "Mysterium Iniquitatis—of Sinful Man Aspiring into the Place of God," in P. G. Frank (ed.), *The Validation of Scientific Theories,* pp. 159–170. Boston: Beacon Press, 1954.
216. MacKay, D. M. "Mentality in Machines," A Symposium, *Aristotelian Society Supplementary Volume,* 26:61–86 (1952).
217. MacLeod, R. B. "The Phenomenological Approach to Social Psychology," *Psychological Review,* 54:193–210 (1947).

218. MacLeod, R. B. "New Psychologies of Yesterday and Today," *Canadian Journal of Psychology*, 3:199–212 (1949).
219. Meehl, P. E. "A Most Peculiar Paradox," *Philosophical Studies*, 1:47–48 (1950).
220. Meehl, P. E. "On the Circularity of the Law of Effect," *Psychological Bulletin*, 47:53–75 (1950). Reprinted as "Law and Convention in Psychology," in H. Feigl and M. Brodbeck (eds.), *Readings in the Philosophy of Science*, pp. 637–659. New York: Appleton-Century-Crofts, 1953.
221. Meehl, P. E., and W. Sellars. "The Concept of Emergence," in H. Feigl and M. Scriven (eds.), *Minnesota Studies in the Philosophy of Science*, Vol. I, pp. 239–252. Minneapolis: Univ. of Minnesota Press, 1956.
222. Mehlberg, H. "Positivisme et Science," *Studia Philosophica*, 3:211–294 (1948).
223. Mellor, W. W. "Three Problems about Other Minds," *Mind*, 65:200–218 (1956).
224. Miller, D. S. "Is Consciousness 'A Type of Behavior'?" *Journal of Philosophy*, 8:322–327 (1911).
225. Miller, D. S. "The Pleasure-Quality and the Pain-Quality Analysable, not Ultimate," *Mind*, 38:150, 215–218 (1929).
226. Miller, D. S. " 'Descartes' Myth' and 'Professor Ryle's Fallacy,' " *Journal of Philosophy*, 48:9 (1951).
227. Miller, J. G. *Unconsciousness*. New York: Wiley, 1942.
228. Morris, C. W. *Six Theories of Mind*. Chicago: Univ. of Chicago Press, 1932.
229. Morris, C. W. *Signs, Language, and Behavior*. New York: Prentice-Hall, 1946.
230. Nagel, E. "The Meaning of Reduction in the Natural Sciences," in R. C. Stauffer (ed.), *Science and Civilization*, pp. 99–138. Madison (Wis.): Univ. of Wisconsin Press, 1949. Reprinted in P. P. Wiener (ed.), *Readings in the Philosophy of Science*, pp. 531–548. New York: Scribner, 1953.
231. Nagel, E. "Are Naturalists Materialists?" *Journal of Philosophy*, 42:515–553 (1945). Reprinted in E. Nagel, *Logic Without Metaphysics*, pp. 19–38. Glencoe (Ill.): The Free Press, 1956.
232. Nagel, E. "Mechanistic Explanation of Organismic Biology," *Philosophical and Phenomenological Research*, 11:3 (1951). Reprinted in S. Hook (ed.), *American Philosophers at Work*, pp. 106–120. New York: Criterion Books, 1956.
233. Nagel, E. *Sovereign Reason*. Glencoe (Ill.): The Free Press, 1954.
234. Nagel, E. "Naturalism Reconsidered," *Proceedings and Addresses of the American Philosophical Association*, 28:5–17 (1954–1955). Reprinted in E. Nagel, *Logic Without Metaphysics*, pp. 3–18. Glencoe (Ill.): The Free Press, 1956.
235. Nagel, E. "A Formalization of Functionalism," in E. Nagel, *Logic Without Metaphysics*, pp. 247–283. Glencoe (Ill.): The Free Press, 1956.
236. Nagel, E., and C. G. Hempel. "Symposium: Problems of Concept and Theory Formation in the Social Sciences." *Science, Language, and the Human Rights*. American Philosophical Association, Vol. I. Philadelphia: Univ. of Pennsylvania Press, 1952.
237. Nelson, E. J. "A Defense of Substance," *Philosophical Review*, 56:491–509 (1947).
238. Nelson, E. J. "The Verification Theory of Meaning," *Philosophical Review*, 63: 182–192 (1954).
239. Northrop, F. S. C. *The Logic of the Sciences and the Humanities*. New York: Macmillan, 1947.
240. Northrop, F. S. C. "The Neurological and Behavioristic Psychological Basis of the Ordering of Society by Means of Ideas," *Science*, 107:411–417 (1948).
241. O'Connor, D. J. "Awareness and Communication," *Journal of Philosophy*, 52: 505–514 (1955).
242. Pap, A. *Elements of Analytic Philosophy*. New York: Macmillan, 1949.

243. Pap, A. "Other Minds and the Principle of Verifiability," *Revue Internationale de Philosophie*, No. 17–18, Fasc. 3–4 (1951).
244. Pap, A. "The Concept of Absolute Emergence," *British Journal for the Philosophy of Science*, 2:8 (1952).
245. Pap, A. "Semantic Analysis and Psycho-Physical Dualism," *Mind*, 61:242 (1952).
246. Pap, A. "Reduction-Sentences and Open Concepts," *Methodos*, 5:17 (1953).
247. Pap, A. "Das Leib-Seele-Problem in der Analytischen Philosophie," *Archiv für Philosophie*, Band 5, Heft 2, 113–129 (1954).
248. Pap, A. *Analytische Erkenntnistheorie*. Vienna: Springer, 1955.
249. Pap, A. "Synonymy, Identity of Concepts and the Paradox of Analysis," *Methodos*, 7:115–128 (1955).
250. Paulsen, F. *Introduction to Philosophy*. New York: H. Holt and Co., 1895.
251. Penelhum, T. "Hume on Personal Identity," *Philosophical Review*, 64:571–589 (1955).
252. Pepper, S. C. "Emergence," *Journal of Philosophy*, 23:241–245 (1926).
253. Pepper, S. C. "A Criticism of a Positivistic Theory of Mind," in G. P. Adams, J. Loewenberg, and S. C. Pepper (eds.), *The Nature of Mind*, pp. 211–232. Berkeley: Univ. of California Press, 1936.
254. Pepper, S. C. "The Issue Over the Facts," in *Meaning and Interpretation*. Berkeley: Univ. of California Press, 1950.
255. Pepper, S. C. "A Neural Identity Theory," in S. Hook (ed.), *Dimensions of Mind*, pp. 37–56. New York: New York Univ. Press, 1960.
256. Place, U. T. "Is Consciousness a Brain Process?" *British Journal of Psychology*, 47:44–50 (1956).
257. Poincaré, H. *The Foundations of Science*. New York: The Science Press, 1929.
258. Popper, K. R. *Logik der Forschung*. Vienna: Springer, 1935.
259. Popper, K. R. "Language and the Body-Mind Problem," *Proceedings of the 11th International Congress of Philosophy*, 7:101–107 (1953).
260. Pratt, C. C. *The Logic of Modern Psychology*. New York: Macmillan, 1939.
261. Pratt, J. B. *Matter and Spirit*. New York: Macmillan, 1922.
262. Pratt, J. B. "The Present Status of the Mind-Body Problem," *The Philosophical Review*, 65:144–156 (1936). Reprinted in *Proceedings and Addresses of the American Philosophical Association*, 9:144–166 (1935).
263. Pratt, J. B. *Personal Realism*. New York: Macmillan, 1937.
264. Price, H. H. *Perception*. London: Methuen, 1932.
265. Prince, M. "The Identification of Mind and Matter," *Philosophical Review*, 13:445–451 (1904).
266. Pumpian-Mindlin, E. (ed.). *Psychoanalysis as a Science* (with essays by E. R. Hilgard, L. S. Kubie, and the editor). Stanford: Stanford Univ. Press, 1952.
267. Putnam, H. "Mathematics and the Existence of Abstract Entities," *Philosophical Studies*, 7:81–88 (1956).
268. Quine, W. V. "On Mental Entities," in the *Proceedings of the American Academy of Arts and Sciences*, 80:3 (1950), Contributions to the Analysis and Synthesis of Knowledge.
269. Quine, W. V. "On What There Is," in W. V. Quine, *From a Logical Point of View*, pp. 1–19. Cambridge (Mass.): Harvard Univ. Press, 1953.
270. Quinton, A. M. "The Problem of Perception," *Mind*, 64:28–51 (1955).
271. Raab, F. V. "Free Will and the Ambiguity of 'Could,'" *Philosophical Review*, 64:60–77 (1955).
272. Rashevsky, N. "Is the Concept of an Organism as a Machine a Useful One?" in P. G. Frank (ed.), *The Validation of Scientific Theories*. Boston: Beacon Press, 1954.

273. Reichenbach, H. *Experience and Prediction*. Chicago: Univ. of Chicago Press, 1938.
274. Reichenbach, H. *Elements of Symbolic Logic*. New York: Macmillan, 1947.
275. Reichenbach, H. *The Rise of Scientific Philosophy*. Berkeley: Univ. of California Press, 1951.
276. Reichenbach, H. "Are Phenomenal Reports Absolutely Certain?" *Philosophical Review*, 61:147–159 (1952).
277. Rescher, N., and P. Oppenheim. "Logical Analysis of Gestalt Concepts," *British Journal for the Philosophy of Science*, 6:89–106 (1955).
278. Rhees, R. "Can There be a Private Language?" A Symposium, *Aristotelian Society Supplementary Volume*, 28:77–94 (1954).
279. Riehl, A. *Introduction to the Theory of Science and Metaphysics*. London: Kegan Paul, Trench and Co., 1894.
280. Roelofs, H. D. "A Case for Dualism and Interaction," *Philosophical and Phenomenological Research*, 15:451–476 (1955).
281. Rosenblueth, A., and N. Wiener. "Behavior, Purpose and Teleology," *Philosophy of Science*, 10:18–24 (1943).
282. Rosenblueth, A., and N. Wiener. "Purposeful and Non-Purposeful Behavior," *Philosophy of Science*, 17:318–326 (1950).
283. Rozeboom, W. R. "Mediation Variables in Scientific Theory," *Psychological Review*, 63:249–264 (1956).
284. Russell, B. *Our Knowledge of the External World*. New York: Norton & Co., 1929.
285. Russell, B. *The Analysis of Matter*. New York: Harcourt, Brace, 1927.
286. Russell, B. *An Inquiry Into Meaning and Truth*. New York: Norton & Co., 1940.
287. Russell, B. *The Analysis of Mind*. New York: Macmillan, 1921.
288. Russell, B. *Human Knowledge*. New York: Simon and Schuster, 1948.
289. Ruyer, R. *Esquisse d'une Philosophie de la Structure*. Paris: F. Alcan, 1930.
290. Ruyer, R. "Les Sensations Sont-elles dans Notre Tête?" *Journal de Psychologie*, 31:555–580 (1934).
291. Ruyer, R. *Elements de Psycho-Biologie*. Paris: Presses Univ. de France, 1946.
292. Ruyer, R. *La Conscience et le Corps*. Paris: Presses Univ. de France, 1950.
293. Ruyer, R. *Néo-Finalisme*. Paris: Presses Univ. de France, 1952.
294. Ryle, G. *The Concept of Mind*. London: Hutchinson's Univ. Libr., 1949.
295. Ryle, G. "The Physical Basis of Mind: A Philosophers' Symposium III," in P. Laslett (ed.), *The Physical Basis of Mind*, pp. 75–79. New York: Macmillan, 1950.
296. Samuel, V. "The Physical Basis of Mind: A Philosophers' Symposium I," in P. Laslett (ed.), *The Physical Basis of Mind*, pp. 65–69. New York: Macmillan, 1950.
297. Scheffler, I. "The New Dualism: Psychological and Physical Terms," *Journal of Philosophy*, 47:25 (1950).
298. Schlick, M. *Allgemeine Erkenntnislehre*. Berlin: Springer, 1925.
299. Schlick, M. *Gesammelte Aufsaetze*. Vienna: Gerold & Co., 1938.
300. Schlick, M. "De la Relation des Notions Psychologiques et les Notions Physiques," *Revue de Synthèse*, 10:5–26 (1935). Reprinted in English in H. Feigl and W. Sellars (eds.), *Readings in Philosophical Analysis*, pp. 393–407. New York: Appleton-Century-Crofts, 1949.
301. Schlick, M. *Problems of Ethics*. New York: Prentice-Hall, 1939.
302. Schlick, M. "Positivism and Realism," *Synthese*, Vol. 7, 6-B (1948–1949) in Communications of the Institute for the Unity of Science, Boston.
303. Schrödinger, E. *What is Life?* Cambridge: Cambridge Univ. Press, 1944.

304. Scriven, M. "The Mechancial Concept of Mind," *Mind*, 62:230–240 (1953).
305. Scriven, M. "Modern Experiments in Telepathy," *Philosophical Review*, 65: 231–251 (1956).
306. Scriven, M. "A Study of Radical Behaviorism," in H. Feigl and M. Scriven (eds.), *Minnesota Studies in the Philosophy of Science*, Vol. I, pp. 88–130. Minneapolis: Univ. of Minnesota Press, 1956.
307. Sellars, R. W. *The Philosophy of Physical Realism*. New York: Macmillan, 1932.
308. Sellars, W. "Realism and the New Way of Words," *Philosophical and Phenomenological Research*, 8:601–634 (1948). Reprinted in H. Feigl and W. Sellars (eds.), *Readings in Philosophical Analysis*, pp. 424–456. New York: Appleton-Century-Crofts, 1949.
309. Sellars, W. "Aristotelian Philosophies of Mind," in V. J. McGill, M. Farber, and R. W. Sellars (eds.), *Philosophy for the Future*, pp. 544–570. New York: Macmillan, 1949.
310. Sellars, W. "Mind, Meaning, and Behavior," *Philosophical Studies*, 3:83–94 (1952).
311. Sellars, W. "A Semantical Solution of the Mind-Body Problem," *Methodos*, 5:45–84 (1953).
312. Sellars, W. "Some Reflections on Language Games," *Philosophy of Science*, 21:204–228 (1954).
313. Sellars, W. "Is There a Synthetic A Priori?" *Philosophy of Science*, 20:121–138 (1953).
314. Sellars, W. "Inference and Meaning," *Mind*, 62:313–338 (1953).
315. Sellars, W. "Empiricism and the Philosophy of Mind," in H. Feigl and M. Scriven (eds.), *Minnesota Studies in the Philosophy of Science*, Vol. I, pp. 253–329. Minneapolis: Univ. of Minnesota Press, 1956.
316. Sellars, W. "Empiricism and Abstract Entities," in P. A. Schilpp (ed.), *The Philosophy of Rudolf Carnap*. LaSalle (Ill.): Open Court Pub. Co., 1963.
317. Seward, J. P. "The Constancy of the I-V: A Critique of Intervening Variables," *Psychological Review*, 62:155–168 (1955).
318. Sheldon, W. H. "Are Naturalists Materialists?" *Journal of Philosophy*, 43:197–209 (1946).
319. Singer, E. A. *Mind as Behavior*. Columbus (Ohio): R. G. Adams and Co., 1924.
320. Skinner, B. F. "The Operational Analysis of Psychological Terms," *Psychological Review*, 52:270–277 (1945). Reprinted in H. Feigl and M. Brodbeck (eds.), *Readings in the Philosophy of Science*, pp. 585–594. New York: Appleton-Century-Crofts, 1953.
321. Skinner, B. F. *Science and Human Behavior*. New York: Macmillan, 1953.
322. Smith, M. B. "The Phenomenological Approach in Personality Theory: Some Critical Remarks," *Journal of Abnormal and Social Psychology*, 45:516–522 (1950).
323. Smythies, J. R. *Analysis of Perception*. London: Routledge & Kegan Paul, 1956.
324. Snygg, D., and Combs, A. W. "The Phenomenological Approach and the Problem of 'Unconscious' Behavior: A Reply to Dr. Smith." *Journal of Abnormal and Social Psychology*, 45:523–528 (1950).
325. Soal, S. C., and F. Bateman. *Modern Experiments in Telepathy*. New Haven: Yale Univ. Press, 1954.
326. Spilsbury, R. J. "Mentality in Machines." A Symposium, *Aristotelian Society Supplementary Volume*, 26:27–60 (1952).
327. Stace, W. T. *Theory of Knowledge and Existence*. Oxford: Oxford Univ. Press, 1932.
328. Stevens, S. S. "Psychology and the Science of Science," *Psychological Bulletin*,

36:221–263 (1939). Reprinted in P. P. Wiener (ed.), *Readings in Philosophy of Science*, pp. 158–184. New York: Scribner, 1953.
329. Stevenson, C. L. *Ethics and Language*. New Haven (Conn.): Yale Univ. Press, 1944.
330. Stout, A. K. "Free Will and Responsibility," *Proceedings of the Aristotelian Society*, 37:213–230 (1937). Reprinted in W. Sellars and J. Hospers (eds.), *Readings in Ethical Theory*, pp. 537–548. New York: Appleton-Century-Crofts, 1952.
331. Stout, G. F. *Mind and Matter*. Cambridge (Eng.): Cambridge University Press, 1931.
332. Strong, C. A. *Essays on the Natural Origin of Mind*. London: Macmillan, 1930.
333. Taylor, R. "Comments on a Mechanistic Conception of Purposefulness," *Philosophy of Science*, 17:310–317 (1950).
334. Taylor, R. "Purposeful and Non-Purposeful Behavior: Rejoinder," *Philosophy of Science*, 17:327–332 (1950).
335. Tolman, E. C. "Psychology versus Immediate Experience," *Philosophy of Science*, 2:356–380 (1935).
336. Tolman, E. C. *Collected Papers in Psychology*. Berkeley: Univ. of California Press, 1951.
337. Tomas, J. "Can We Know the Contents of C. I. Lewis' Mind?" *Philosophical and Phenomenological Research*, 11:541–548 (1951).
338. Turing, A. M. "Computing Machinery and Intelligence," *Mind*, 59:433–460 (1950).
339. University of California Associates. "The Freedom of the Will," in *Knowledge and Society*. New York: D. Appleton-Century Co., 1938. Reprinted in H. Feigl and W. Sellars (eds.), *Readings in Philosophical Analysis*, pp. 594–615. New York: Appleton-Century-Crofts, 1949.
340. Wallraff, C. F. "On Immediacy and the Contemporary Dogma of Sense-Certainty," *Journal of Philosophy*, 50:2 (1953).
341. Watling, J. "Ayer on Other Minds," *Theoria*, 20:175–180 (1954).
342. Watling, J. "Inference from the Known to the Unknown," *Proceedings of the Aristotelian Society*, New Series 55:83–108 (1955).
343. Weber, A. O. "Gestalttheorie and the Theory of Relations," *Journal of Philosophy*, 35:589–606 (1938).
344. Weber, C. O. "Theoretical and Experimental Difficulties of Modern Psychology with the Body-Mind Problem," in P. L. Harriman (ed.), *Twentieth Century Psychology*, pp. 64–93. New York: Philosophical Library, 1946.
345. Weitz, M. "Professor Ryle's 'Logical Behaviorism,'" *Journal of Philosophy*, 48:9 (1951).
346. Wenzl, A. *Das Leib-Seele-Problem*. Leipzig: Verlag von Felix Meiner, 1933.
347. White, M. *Toward Reunion in Philosophy*. Cambridge (Mass.): Harvard Univ. Press, 1956.
348. Wiener, N. *Cybernetics*. New York: Wiley, 1948.
349. Williams, D. C. "Scientific Method and the Existence of Consciousness," *Psychological Review*, 41:461–479 (1934).
350. Wilson, N. L. "Designation and Description," *Journal of Philosophy*, 50:369–383 (1953).
351. Wilson, N. L. "In Defense of Proper Names Against Descriptions," *Philosophical Studies*, 4:73–78 (1953).
352. Wilson, N. L. "Property Designation and Description," *Philosophical Review*, 64:389–404 (1955).
353. Wisdom, J. "Other Minds," *Aristotelian Society Supplementary Volume*, 20:122–147 (1946).

354. Wisdom, J. *Other Minds*, Oxford: Blackwell, 1952.
355. Wisdom, J. O. "Mentality in Machines," A Symposium, *Aristotelian Society Supplementary Volume*, 26:1–26 (1952).
356. Wisdom, J. O. "Is Epiphenomenalism Refutable?" *Proceedings of the 2nd International Congress of the International Union for the Philosophy of Science*, 5:73–78 (1954).
357. Wittgenstein, L. *Philosophical Investigations*. London and New York: Macmillan, 1953.
358. Woodger, J. H. *Biological Principles*. London: Kegan Paul, Trench & Co., 1929.
359. Woodger, J. H. *Physics, Psychology and Medicine*. Cambridge: Cambridge Univ. Press, 1956.

POSTSCRIPT AFTER TEN YEARS

Postscript after Ten Years

The foregoing essay was completed in February 1957, and published in the spring of 1958. During the last ten years, and especially during the last five or six years, I have received a veritable avalanche of extremely stimulating responses. Some of these are contained in books, others in essays and articles, and there have also been manuscripts, Ph.D. theses here at the University of Minnesota and elsewhere (some of them still unpublished at the moment), as well as a great many letters. I have kept, and still keep, trying to react to this "feedback" as fully as I can by way of correspondence. I have also had the benefit of critical discussions by philosophers and scientists on scores of occasions —lectures, symposia, colloquia, seminars, etc., both at the University of Minnesota and elsewhere in the United States, in Austria (1964), and in Australia (1965). Many of these new contributions toward a discussion of one or the other facet of the mind-body problem referred directly to one or several of my publications; others deal with the closely related but different views of J. J. C. Smart, or other philosophers or scientists. I am most grateful for all these—widely differing —responses. They ranged from almost complete agreement to incisive constructive as well as destructive criticism. In the brief remarks that follow, I shall be able to discuss only a limited number of these published or unpublished views and criticisms. I shall restrict myself to what appear to me the most important points. This, of course, involves the risk that I may overlook (but I hope not "suppress" or "repress") some essential criticisms. There are new approaches now in the making and in need of fuller development. With these I can deal only very sketchily, mainly because I have not been able fully to understand them, let alone to appraise them adequately and accurately.

In any case I feel somewhat vindicated in my view that the mind-body problems cannot simply be made to disappear by purely linguistic maneuvers (60). The fashions of philosophy change, and it seems the mind-body problems are once again in the forefront of highly active and intelligent philosophical discussions.

To begin with, let me frankly state that despite all the extant contributions toward a solution or even only a precise clarification of the several puzzles that constitute the perplexities of mind-body problems, I am not aware of any solution that is completely successful. My own extensive essay* written in 1957 now appears to me questionable in many points. I mentioned and discussed several of these vulnerable and problematic points quite candidly in section VI ("A Budget of Unsolved Problems. Suggestions for Further Analyses and Research").

As I see it now more clearly than before there are unresolved difficulties in each of the three main issues—be they scientific or conceptual-philosophical—of the cluster of mind-body problems: *sentience*, *sapience*, and *selfhood*. The sentience problem was most succinctly and elegantly formulated by Mrs. Judith Economos in a preliminary way at the very beginning of her doctoral dissertation (completed early in 1967 at the University of California at Los Angeles). With her kind permission I quote fairly fully (making only very slight stylistic changes):

> . . . I shall first try to exhibit the existence of a perplexity, which I take to be the mind-body problem. This I shall do by listing four propositions which appear to be true but are difficult to reconcile with each other. Then I shall sketch various proposed solutions to the mind-body problem as elicited from the four propositions, and try to show that each proposed solution appears to conflict with one of the four propositions. . . .
>
> Four Propositions
>
> 1. People have sensations, thoughts, etc., of which they are aware, but of which others are not aware except through the owners' reports or behavior; and these sensations, thoughts, etc. are not located in space, nor do they possess, produce, or consume energy or have mass.
>
> 2. There are in the world various objects, including contemporary human bodies, which are composed of elements which are located in space, which produce, possess, and consume energy and have mass; moreover the vocabulary of physical science seems sufficient to describe,

* Designated most flatteringly by John Beloff (19) as the "Russell-Feigl Identity Theory," and characterized—also by Beloff (171)—very humorously as "whatever is the opposite of a nutshell"; I admit it was perhaps "a little long for its length"!

and the laws of physical science seem sufficient to explain, the behavior of such objects.

3. People's sensations, thoughts, etc., affect their bodily movements, and some events occurring to or in their bodies affect the people's sensations, thoughts, memories, etc.; and often when people's sensations, thoughts, etc., affect their bodily movements, this is because the people have so desired.

4. The concepts which we have of mental things or events on the one hand, and of material things or events on the other, are logically independent; that is to say, there is no demonstrable inconsistency in supposing a world in which there were material objects but no awareness, or conversely a world of awarenesses without anything in it fitting the description of matter.*

At first glance it would appear that these incompatible four propositions force us willy-nilly into the agnostic position expressed so poignantly last century by the German physiologist E. Du Bois-Reymond in his famous phrase "*ignoramus et ignorabimus*"; but most philosophers remained undaunted by this "riddle of the universe" and have tried—in extremely diverse ways—to solve or to "dissolve" it. I have dealt briefly with the major forms of "dissolution": Radical Materialism; Physicalism (of various types); Logical Behaviorism; Neutral Monism; Phenomenalism; Oxford Linguistic Analysis; etc. As I have pointed out (60), despite the powerful impressions made upon me in my Vienna Circle years (1924–30), I no longer consider *most* mind-body puzzles as pseudoproblems engendered by conceptual confusions. I rather think—and have thought so at least for the last thirty-five years—that the relation of the mental to the physical presents us with a cluster of genuine and complex problems: some primarily scientific, others primarily philosophical. Perhaps I should mention immediately that I have come to think with increasing conviction that there is no sharp line of demarcation between (good) science and (clearheaded) philosophy. Every major scientific advance involves revisions of our conceptual frameworks; and doing philosophy in our day and age without regard to the problems and results of the sciences is—to put it mildly—intellectually unprofitable, if not irresponsible.

Once the pessimistic-agnostic position is abandoned, one faces the arduous task of working out a truly synoptic solution that is logically consistent. Consistency in these matters is painfully difficult to achieve

* In her thesis, Mrs. Economos deals later on also, and I think in part extremely effectively, with the problems of sapience and especially of selfhood.

—as I know only too well from my own experience in groping for an all around satisfactory solution. In regard to the problem of sentience this is readily evident. Inasmuch as one wants to retain the essential contentions of physicalism and to repudiate epiphenomenalism (as is, for example, the basic tendency of J. J. C. Smart and other members of the "United Front of Sophisticated Australian Materialists" such as D. M. Armstrong, Brian Medlin, R. Routley), one embraces a "central state theory of mind." This amounts to the claim that—"in principle"—a physical description of the world is *complete*, i.e., leaves out nothing. This, obviously, takes some doing; where or how are the apparently "homeless" qualities of immediate experience to be located? The first step is, of course, that taken by Bertrand Russell (as in *Human Knowledge*). The "homeless" qualities are "really in the mind"; and being mental they are features of cerebral processes. But what sort of features? In keeping with the best evidence of psycho-neurophysiology, they are very likely "global"—i.e., configurational, Gestalt-like features of the much more "finely grained" neural processes (and, a fortiori, of their "micro-micro"—i.e., atomic, subatomic, and quantal—constituents).

One challenging way of pointing up the issue is to ask whether the physicalistic account can *really* be "complete." I had the privilege of discussing the problem (along with many more "tangible" matters in the philosophy of physics) with Albert Einstein one afternoon in April 1954 at his home in Princeton. I asked Einstein whether in an ideally perfect (of course utopian) four-dimensional, physical representation (à la Minkowski) of the universe the qualities of immediate experience (we called them metaphorically the "internal illumination" of the "knotty clusters of world lines" representing living-awake brains) were not left out. He replied in his characteristic, humorous manner (I translate from the German in which he used a rather uncouth word): "Why, if it weren't for this 'internal illumination' [i.e., sentience] the world would be nothing but a pile of dirt!" This reply suggests that the (ultimate-utopian) physicalistic account, though complete in one way, is incomplete in another. But to put this in an enlightening and consistent manner is precisely the most thorny and important task for the identity theory.

I had hoped that my own double-knowledge, double-designation view would yield what is wanted. This view would retain the basically empirical (synthetic) character of the mode of ascertaining the identity. Then, if the identity were assumed, the objectionable feature of epiphenomenalism would be eliminated. The arguments from the "causal efficacy" of pleasure, displeasure, attention, vigilance, desire, and volition, which the present-day Neo-cartesians* keep using for interactionism, do not—I think—refute the identity theory. (I shall discuss the related arguments from intentionality and the unity of consciousness when I come to deal with sapience and selfhood.) The familiar use of interactionist statements in ordinary language, though logically delicate and hazardous, is not in principle objectionable. It is one and the same event, say a decision or volition, or a sudden pain, described phenomenally in one way, and physically in another way, which is a causal antecedent of a "bodily" response or movement; or, vice versa, some physical stimulus input causes a central state—described either in the familiar phenomenal language as a sensation, or in the (utopian) physical language as a feature of a cerebral process.

Now, the crucial issue is: Does my form of the identity hypothesis involve the assumption of "nomological danglers"? In many discussions my great and good friend Professor J. J. C. Smart attempted to show me that I don't need the "danglers." But I was recalcitrant in that I didn't see how one could maintain the empirical (synthetic) character of the identification which Smart, in his own way, also stresses, and which must in some way reflect the correlations and isomorphisms that are gradually and increasingly discovered by psycho-(neuro)-physiology. Yet, right from the beginnings of my reflections on the traditional puzzle I was convinced that the "danglers" are metaphysically quite innocuous. Smart, I think, is essentially right in that they would not, and could not, appear in the "finished" scientific conception of the world. This can be made plausible by considering once again the piece of science fiction—the Martian Super-Scientist. Let us assume that a complete explanation of animal and human behavior can be achieved by reduction to the basic physical laws, and that the structures (initial and boundary conditions) of organisms can be described in purely $physical_2$ terms; then there is no need for the phenomenal terms —just as there would be no need for typically biological or physiologi-

* Beloff, Ducasse, Shaffer, Popper, and Chisholm—in their various ways—renew on a more sophisticated level the old arguments of McDougall, Bergson, and Driesch.

cal concepts. They would all be "reduced" to whatever are the concepts of the "ultimate" physics (e.g., something like the concepts of current atomic, quantum, and field physics). The Martian's repertory —if he has a repertory of qualities of immediate experience at all (i.e., if he is not a "mere robot")—may not in any way overlap with that of us earthlings. In that case he would lack altogether any "acquaintance" with the qualities of our "raw feels." He would consequently also lack the sort of "empathy" that humans can have for each other. The physicalist would formulate this, of course, by pointing to essential differences between the Martian and the human central states and processes. The Martian would thus not know "what colors look like"; "what musical tones sound like"; "what joy, grief, elation or depression, etc., etc. feel like." Nevertheless he would be able to explain—and possibly also to predict—all of human behavior on the basis of his micro-theories. His theories may be expressed in a notation (reflecting concept formation) utterly different from our basic physics—but his physics would nevertheless be completely translatable into ours and vice versa.

Now, the question arises: Is there something about human beings that the Martian does not (and never could) "know"? This is merely the question of "completeness" over again, of course. I think Paul E. Meehl is correct in arguing (115) that the possession of a certain set of raw feels implies a cognitive advantage—in that a Martian (now of a different kind) would be in a position to explain and predict human behavior much more readily if his repertory of direct experience at least partially overlapped with ours. But I maintain that, given enough time and intelligence, the Martian with even a totally different repertory of raw feels would in principle (although much more cumbersomely and slowly) arrive at a complete explanation of the behavior of the earthlings.

I am inclined to think that the basic philosophical issue lies elsewhere. I believe that those thinkers who maintain that a "category mistake" is involved in mixing phenomenal and physical language are essentially right. The only trouble is that we have thus far had no precise and convincing explication of the very notion of a category mistake (of this kind!). There are, of course, other sorts of category mistakes, e.g., the mistakes that arise out of a confusion of Russellian-type levels. I don't think that the theory of types is relevant here. The initially suggestive ideas along these lines in R. Carnap's *Der Logische Aufbau der Welt* (1928, now available in English translation) or in G. Ryle's *The Concept of Mind* (1949) no longer seem adequate; the simple reason being that neither the phenomenalistic (or neutral-monistic) reconstruction of the *Aufbau* nor the logical-linguistic behaviorism of Ryle is acceptable in the light of recent criticisms.

I, too, have to admit that the special formulation I presented in the long essay ten years ago must be revised. I did see that the sort of identification that seems legitimate and plausible (although also open to logico-methodological criticisms) in the natural sciences (e.g., temperature = mean kinetic energy of molecules; table salt = NaCl; Mendel's factors = genes containing DNA and RNA; and the like) can serve only if we conceive of psychology as a branch of biology, i.e., if we adopt the conceptual frame of *intersubjective* science. In that case such identifications as (short-term) memory trace = reverberating neural circuit, attention and vigilance = activation of the reticular formation in the brain, etc., etc. are logically and methodologically on a par with those mentioned before; or to give one more example, the identification of ferro-magnetism with a certain (statistical) distribution of the spin of electrons in the iron atoms.

Even this way of formulating the identifications has been called into doubt by Carnap and Feyerabend. These two philosophers of science, while perhaps allowing that "identification" is appropriate in its basic intent, maintain that it must really be regarded as an *explicandum*. The full and more accurate *explication* of it should be rendered in terms of *fusion* (Carnap, personal communication) or, similarly, as *replacement* or *supplantation* of concepts at earlier stages of scientific theories by concepts of a later, more accurate, and more comprehensive theory (67, 68, 69, 70). This is, however, not the place to enter into a detailed discussion of this important issue.

In any case, I now agree with Smart (and perhaps with Feyerabend) that within the conceptual frame of theoretical natural science genuinely phenomenal (raw feel) terms have no place. Although the following analogy is almost sure to mislead, I shall nevertheless use it as a "bridge" toward the denouement I am going to suggest a little later. My point is this: just as the commonsense (direct-realist) concepts of surface color, tone quality, flower fragrance, heat intensity, tangible hardness, etc. are supplanted by their "successor concepts" (a felicitous term used by W. Sellars) in physical theory, such as frequency of electromagnetic waves, frequency, etc. of acoustical waves, chemical structure of "aromatic" compounds, molecular motion, atomic structure (of, for example, the diamond), etc.—so the phenomenal predicates used in the description of after-images, sensations, feelings, emotions, moods, etc. are to be replaced by the (as yet only sketchily known) neurophysiological and ultimately micro-physical characterizations.

But just as (good) science never "explains anything away" (except the "objects" of superstitions, illusions, or hallucinations), so the phenomena of the world of common experience (be they external, i.e., extradermal; or internal, i.e., intradermal; or "internal" in that other—tricky—sense of "mental") are *explained*, but not explained away. The "successor concepts" may be, and usually are, far removed from their "predecessors"; they don't have the warm familiarity, the colorful, "Christmasy" pictorial and emotional appeals of the commonsense terms; but they have far greater explanatory power and coherence.

Wilfrid Sellars (155, 156) has offered a highly suggestive analysis along these lines by distinguishing the "manifest image" from the "scientific image" of the world. In keeping with what I have said in the long essay about the meaning of physical concepts, I would prefer to contrast the manifest image with the scientific *conception* of the world. More strongly than ever before, I am convinced that it is primarily the concept of the "physical" that requires reinterpretation and reconstruction. The imagery that is so helpful heuristically and didactically is not and cannot be part of the cognitive meaning of physical concepts and hypotheses. If we construe physical theories with the help of Ramsey sentences as Carnap proposes (32), then it becomes clear that our knowledge of the "physical world" is "purely structural." That is to say that the postulates of physical theories, including the correspondence rules, give us only an "implicit" definition of the theoretical concepts of the physical sciences. And since the concepts of biology and psychology (if physicalism is correct!) are reducible to those of physics, the same holds for the concepts of all behavioral sciences. Russell and Schlick were essentially right in saying that we have merely knowledge by ("structural") description of the "physical world." What then, by contrast, is "knowledge by acquaintance" as we have it "subjectively" and introspectively? Here I still hold in essence the view formulated ten years ago: Since all knowledge is propositional, the propositions that formulate knowledge by acquaintance can do no more than reflect the *structure* of whatever is "given in immediate experience."

At this point once again arises the perplexity of the "ineffable" qualia of direct experience. Poincaré, Eddington, and other brilliant scientist-philosophers have made much of the distinction between the "inexpressible" and "uncommunicable" *contents* and the propositionally expressible and communicable *forms* or *structures* of the immediately given. As before, I believe this is a highly suggestive but nonetheless extremely misleading formulation. Schlick was aware of these dangers, but was not quite able to avoid them himself. I shall now try to explicate the distinction as well as I can. The first and perhaps most important point to notice is the essential difference between the concepts of the physical sciences and the concepts of introspective-phenomenological psychology. The concepts of physics may be said to be independent of, or invariant with respect to, their specific "anchoring" in the qualities or modalities of immediate experience. This can be illustrated by the case of congenitally blind persons who would "in principle" be able to arrive at the *same* theoretical physics, astronomy, chemistry, biology, and behavioral psychology as that achieved by persons blessed with eyesight. Given modern electronic devices (photoelectric cells, spectroscopes, transducers, amplifiers, radios, etc.) and enough time and intelligence the blind man could hear with his earphones or from a loudspeaker certain sounds that would lead him to essentially the same conception of the world that is embodied in our science textbooks. *Mutatis mutandis*, this same sort of philosophical "science fiction" could be spun out for a science on the basis of touch, smell, or heat sensations. As long as whatever exists and occurs in the world can in some way be causally connected by special devices with one or another of our sense modalities, and if the discrimination can be made sufficiently sensitive with the help of these devices, it does not matter which data of direct experience serve as the "observables" (in which the correspondence rules "anchor" the theoretical concepts of science). Hence, the "meaning" of physical concepts is invariant with respect to such "transformations" of the observation basis.

The "meaning" of purely phenomenal concepts, such as 'red', 'warm', 'sad', 'glad', by contrast, is quite different—I am inclined to say it is even a different type of meaning (if not meaning of "meaning"). Within the confines of the purely subjective, introspective, phenomenal perspective, there is no such invariance with respect to modality transformations. This is what I meant by saying that purely phenomenal terms are "mere labels" of the qualities they designate. That we have acquired the labeling dispositions through the learning of language (in the way characterized by Carnap, Ryle, and Skinner)* is admitted, but irrelevant to my philosophical point: once we *have* the labeling ability, there is *one* meaning, or rather *type* of meaning (or significance) of phenomenal terms that is radically different from that of physical concepts. Their designata are confined to the range of actual and possible data of direct experience and their immediately given qualities and relations. While the concepts of the (intersubjective) physical sciences, in order to have empirical significance, must of course be "anchored" (by correspondence rules) in the phenomenally given, their meaning is "structural" and non-intuitive in that it involves essentially ("implicit") specification by postulates.

The consequences of all these considerations for the identity theory, as far as I can see, are as follows: Inasmuch as a good and complete physicalistic (i.e., $physical_2$) account of the world will contain "successor" concepts to all phenomenal concepts, there will indeed be no "nomological danglers" in such an account. Nothing important is omitted in such a description; but, of course, what counts as "important" are the spatio-temporal-causal features that are essential for the world's description, explanation, prediction, and retrodiction (as much as whatever degree of fundamental determinism or statistical regularity permits). Even the "anchoring in the data" is represented, but, of course, not in the sense in which traditional epistemology (including my own account above) is accustomed to put it. In the scientific conception of the world, theories of perception, of learning, and of language, ultimately formulated in $physical_2$ concepts, become the "successors" to the phenomenological-epistemological account. This is essentially what is tenable and defensible in modern physicalism (and Australian materialism).

For many years I opposed materialism, holding that it is illegitimate-

* Emphasized in the oft-repeated Wittgensteinian arguments against the possibility of a purely private language. I still consider these arguments as invalid or confused (see 37, 45, 118, 175, 184).

ly reductionistic. That is why I attempted to replace it by my version of the identity theory. I felt that not only the radical behaviorists, but also the materialists somehow suppressed the "other perspective"; that they practiced what I called the "Hylas touch"—i.e., equipped with their particular sort of "blinkers" they turned whatever they touched into "matter" or physical events and processes. But the very possibility of giving a complete $physical_2$ account of the world is just that striking (and logically contingent but basic) feature of the universe and man's place in it that the advancing sciences make increasingly plausible. Nothing is "explained away"—everything is merely being encompassed by a comprehensive conceptual system, no matter how unfamiliar its pivotal concepts may be.

Once again we must ask: What precisely then happens to the familiar phenomenal features of the world as we know it in everyday life? And we answer—first sketchily and metaphorically: they are replaced, transformed, supplanted by the more rigorous, consistent, and explanatorily more coherent and fruitful features of the world as represented by $physical_2$ concepts. Much work still needs to be done toward a full analysis and clarification of this "great transformation." I realize that my own previous ("identity") account must also be thoroughly revised. As it stood ten years ago, it contained insuperable difficulties, particularly in view of the stringent demands of Leibniz's definition of identity in terms of indiscernibility. If I had been satisfied with merely *extensional* identity, I would have been saddled with an ontology of particulars (preferably of events) with *dual properties*. But that is hardly a step in the direction of the thoroughgoing monism I hoped to vindicate.

As I see it now, all *purely* phenomenal statements contain *egocentric universals* (i.e., words designating purely experiential qualia) and many such statements contain, in addition, also *egocentric particulars* (i.e., words like 'this', 'I', 'now', 'here', and/or cognate expressions). The contributions of N. Goodman, H. Reichenbach, B. Russell, Y. Bar-Hillel, W. Sellars, *et al.* to the analysis of 'egocentric', 'token-reflexive', or 'indexical' terms in their pragmatic contexts are important in this connection. Thus far it seems only the egocentric *particulars* have received the attention of logicians and language analysts. It has been shown that the very link of the intersubjective language with the experience of the "knowing subjects" who use that language is given by the pragmatic context of their utterances. The uniqueness of reference of the indexical terms is explicated in the intersubjective frame of science by definite descriptions (unique characterizations) of the moment of utterance, the speaker who produces the utterance, or the place in which the utterance occurs (the word 'occurs' in the last sentence is to be understood in the sense of the "timeless present"). Let me explain this just a little more fully. In order to understand a sentence containing a temporal designation (a date in history for example), I have got to know where in time my "present" experience occurs. In order to understand, for example, geographical or astronomical place designations, I must know "where I am at" in space. (This is brought out humorously by the absurd story of the Boy Scout who on a long hike said to one of his companions: "According to this map we ought to be on that mountain over yonder!") Unless we can locate ourselves on (or in) the "map" of the Minkowski world, we would never understand any place or date designations. But in the intersubjective (Minkowski) representation, the 'here' becomes just one place among indefinitely many others; the 'I' one person (or organism) among others; the 'now' one moment among others. In this transformation ("democratization") the "existentially poignant uniqueness" of the NOW, the HERE, and the I are lost, because they are replaced by such definite descriptions as "the date on which H. Feigl got his Ph.D. degree"; "the place in which the tornado of 1965 did the damage"; "the person who was hit by a meteorite on December 12, 1954." "Uniqueness" in ordinary and scientific contexts simply amounts to a singularity that is logically contingent, but may be empirically demonstrable, or at least plausible. The "existentially poignant uniqueness" of the NOW, the HERE, and the I is a matter of immediate experience. The "successor terms" in the language of science are experientially neutral; they do not have the emotive (i.e., pictorial and emotional) appeal of the phenomenally given significance of the terms as understood, for example, in the "existential anguish" of a life situation, such as when I say to myself, "NOW is the moment to make my decision"; "HERE I shall build a house"; "I alone bear full responsibility for THIS action."

If tough-minded positivists fail (or refuse) to understand this "existentially poignant" uniqueness, there is little that I can do to help them. Only by some sort of arguments *ad hominem*, combined with ostensive procedures, can I convey what I mean. In any case, I can reassure the positivists that I have not the slightest inclination to develop an existentialist metaphysics à la Heidegger. Nor do I subscribe to Wittgenstein's ineffability doctrine ("whereof we cannot speak, thereof we must be silent"—after all I have just spoken, and I hope intelligibly, about matters he thought one could at most "stammer" about).

A rigorous explication of the role of indexical terms should be provided in the semiotic (metalinguistic) discipline of pure pragmatics. But if this is going to be analogous to the explications of pure syntax and pure semantics, it will have to be formulated in an intersubjectively intelligible metalanguage; and hence again the "existential uniqueness" will be relegated to the limbo of emotive significance and supplanted by the neutral "sober and colorless" objective characterization.

Now, while I think that a world description (à la Minkowski) can be given that is—necessarily—devoid of indexical terms, such a world description can neither be fully understood nor practically used without being *linked*—with the help of indexical terms—to the experience of a sentient and sapient (i.e., human) being. This becomes evident if the Minkowski representation is viewed as a map of "all there is" in space-time. If I am to find the "picture" of myself-at-a-given-time on this map, I would have to scrutinize it in its (possibly) infinite extent in order to find just that particular skein (or segment of the set) of world lines which uniquely characterizes me-at-that-time. (If I had an exact double, this procedure would fail.) In actual practice I would, of course, *point* to that small region of the map. This is one way of illustrating the use of indexical terms—and of avoiding the paradox of the Boy Scouts.

It seems to me that what holds of indexical (or egocentric) particulars holds—*mutatis mutandis*—analogously of indexical (egocentric) universals. I cannot even begin to "get a public language going" unless I understand the private (egocentric) language whose predicates (monadic, dyadic, etc.) designate experiential qualities or relations. I must be able to know (by "acquaintance") some phenomenal qualities and relations (redness, between-ness, etc.) in order to "hook" (i.e., connect) my private language to the intersubjective language of science. To the extent that, for example, pointer readings belong to the confirming or disconfirming evidential data of physics, I must be able to "recognize" the position of a pointer on a scale "when I see it." In my proposed reconstruction it is my private impressions, e.g., the shapes and colors in my visual field, which constitute "ultimate" data of observation. I realize that I shall meet here with a storm of opposition because all this will appear to be a restatement of the much-criticized doctrine of sense data. But although I definitely reject the phenomenalistic reduction of physical-object statements to sense-data statements, I must say that I am not in the least impressed by the ordinary language arguments regarding the common use of such words as "observing", "seeing", "hearing". I would argue that these words are not always used as success words or achievement words even in ordinary language. If I close my eyes and press with my fingers on my eyelids I "see" kaleidoscopically changing patterns of colors; but I don't "see" (in the achievement sense) an external physical object. If I have the familiar experience of "ringing in my ears" (i.e., that kind of "hearing a sound"), this may well be no perception of a distant bell, but an experience engendered by intradermal events.

I trust it is clear that I am not for a moment endorsing any doctrine of phenomenalism. I do not even wish to defend the notion of a full-fledged phenomenal language. I merely maintain that by giving ourselves a sort of "wrench" (away from the normal life perspective, probably somewhat similar to what Husserl meant by "bracketing out" all the usual and mostly automatic interpretations and/or inferences) we can arrive at the "given". I cannot see that the "given" in this sense is a myth; but I admit it usually is a "reduct" or "destruct" of a much fuller experience that involves a good deal of conceptual structure or "implicit knowledge". I also admit, and would even stress, that whatever we can say about the given qualia is "structural" at least in the sense that *such* "knowledge by acquaintance" involves much more than the *having* (i.e., undergoing, enjoying or suffering, living through) of an experience. The mere classification of the experientially given in regard to qualities and modalities requires at least the sort of conceptual structure that is constituted by the system of similarities and dissimilarities, and the degrees thereof, as, for example, represented by the topological ordering of (experiential) colors in the well-known color octahedron.

Some thirty-five years ago, i.e., in the heyday of positivism, I would have said that the meaning of purely phenomenal terms is emotive (pictorial, emotional) and non-cognitive. I would have said that this type of significance is exclusively that of expression and evocation. I no longer hold this view. The ostensive link in the "anchoring" of all our empirical concepts very definitely fulfills a cognitive function. Moreover, far from being mere "barkings" (i.e., expressions such as "pain behavior" when crying 'ouch!') or "avowals" (in Ryle's sense), phenomenal descriptions of momentary direct experience do make truth claims, even if their truth is not establishable by "criteria" in the usual sense. They represent the extreme lower limit of cognition; they constitute, admittedly, a "degenerate" and "highly impoverished" sort of knowledge. Nevertheless, they are the "ultimate" basis of all our empirical knowledge claims. It is in this sense, and in this sense only, that I countenance a "methodologically solipsistic" (or "egocentric") reconstruction. The data of direct experience provide the ultimate confirming or disconfirming evidence of all our factual knowledge. Purely phenomenal assertions require no other evidence than that which is "given"; I would call them "self-evident" if this phrase had not been badly misused in traditional epistemology. Of course, as *assertions* (spoken, written—symbolized in any form, even if only "silently thought") I don't consider them infallible ("incorrigible"), for "there is many a slip between the brain and the lip." I even insist on their corrigibility in the wider context of intersubjective discourse and knowledge. Yet in this (solipsistic) reconstruction they are the least dubitable knowledge claims on which any more ambitious knowledge claims (of commonsense and of the factual sciences) are based—"in the last analysis"!

Moreover, if the sort of structuralism and physicalism discussed above holds, then—to express it first in the more familiar dualistic way—the configurational (Gestalt) features of immediate experience are isomorphic with certain global features of our brain processes. Hence, strange as it may sound at first, it is possible that by doing introspective-phenomenological description of immediate experience, we are in effect (though we are hardly ever aware of it) doing also a bit of (very crude, vague, and preliminary) brain physiology. This is my current reply to the "sticky" question: "How does an identity theorist explain the fact that he can worry about the place and role of the raw feels if they are to be identical with brain processes?" While I would no longer speak strictly of "identity" (for reasons discussed above), my answer would simply be that the scientifically uninformed person, when giving phenomenological descriptions, does not *know* that he is at the same time describing certain features of his brain processes. This is to be viewed as a case of what Quine calls "referential opacity." It is in some respects analogous to, for example, the case of the housewife who by saying "the soup is hot now" does not know that she is referring to a state of the soup which (in the light of the modern theory of heat) is characterized also by the mean kinetic energy of the molecules that are the constituents of the soup. In all these cases there is certainly at least identity of *reference* (extension); and there is also identity of some (but not all) *structural* properties (intension). The intensional identity concerns the isomorphism (sameness of structure) of certain global, i.e., statistical and/or Gestalt, features of the micro-states with the more directly observable features of the macro-states.

As is fairly generally agreed, the purely epistemic features ("known", "unknown", "believed", "not-believed", "doubted", etc.) are "intensional" in the narrower sense, in that the *salva veritate*, let alone the *salva necessitate*, condition for substitutions is not required for the usual nomological or systemic (theoretical) and in *that* sense intensional identities.

I still agree, of course, with Wilfrid Sellars (154), Roderick Chisholm (36), Stephan Körner (100), and others, in considering clearly *intentional* (in Brentano's sense) features as irreducible to a physicalistic description. But as I have briefly indicated in the long essay, this does not seem to me a serious flaw in physicalism. According to Sellars' decisive analysis, this irreducibility is on a par with (if not a special case of) the irreducibility of logical categories to psychological or physiological ones. Logical categories, and principles formulated in terms of them, are indeed "categorially" different from those of the factual sciences. Logic (syntax and semantics) is, of course, indispensable in the object language or the metalanguage of all sciences (formal or factual); but the difference between logic and psychology is just as fundamental as that between, say, logic and physics. To disregard the difference amounts to making one of the most glaring category mistakes. This sort of category mistake is fundamentally different from (a) violations of Russell's type rule; (b) confusions of language levels, e.g., object language and metalanguage; (c) mixing of phenomenal with strictly physical concepts; (d) confusion of dispositions (capacities, propensities) with occurrences (episodes, events, processes); (e) mistaking purely emotive (i.e., pictorial, emotional, and/or motivative) expressions and appeals for cognitively meaningful sentences. The "naturalistic fallacy" (i.e., the alleged inference from 'is' to 'ought') is an important example of this categorial confusion. (So here we have then at least six radically diverse kinds of category mistakes; perhaps there are still many others! ?)

The foregoing notwithstanding, some important qualifications are in order. As Paul Meehl* has shown quite cogently, certain logical categories are indispensable in the *molar*-psychological accounts of linguistic (generally, of symbolic) behavior. Consider, for example, the recognition of a piece of reasoning as a fallacy of four terms (*quaternio terminorum*). It is impossible to give a purely physicalistic characterization of the conceivably unlimited varieties of stimulus patterns that would form the class of this kind of fallacy. The stimulus objects might be visual (as in writing or print); they might consist of spoken sounds, of Morse code clicks; of smoke signals, of the gestures of deaf-mute persons; etc., etc. Hence in a *molar*-psychological account, the defining characteristic of the many and varied stimulus patterns that might elicit the response "fallacy of four terms" can be given only in terms of *logical* categories. But, of course, if we had the ideal (utopian) neurophysiological, or ultimately microphysical, description of the cerebral processes that occur in the behavior, for example, of a logic teacher, the precise response would become predictable (at least to the extent and to whatever degree determinism holds in this domain) on the basis of a purely physical description of all the details of the stimulus input, cerebral transaction, and response output; hence the logical categories would then not be required for a characterization of the stimulus classes. But of course the ultimate physical processes would be quite "opaque" to one who could not provide such a classification. This would be analogous to, let us say, a prediction of the configuration of musical notes put on paper by a composer whose "output" is predicted merely in physicalistic terms and would thus be unintelligible to someone not familiar with the rules of musical notation.

These qualifications should, however, not be misunderstood. Whatever occurs in the "mind" of a logician, mathematician, inventor, com-

* In essays to be published in due course, one in the forthcoming Vol. IV of the *Minnesota Studies in the Philosophy of Science*.

poser, etc. is of the nature of a *process*, and hence—if physicalism holds—would be describable and explainable in terms of micro-concepts and laws. For the *molar* psychologist it is again essential to utilize the concept of rule-governed behavior. This has happily come to the forefront of the investigations pursued by the psycholinguists (Chomsky, Fodor, Katz, *et al.*). Obviously, first-level rule-conforming behavior must be distinguished from such second-level rule-governed behavior that amounts to an articulate *statement* of the rules. The ordinary processes of deductive inferences, for example, may well go on without an explicit awareness of the rules to which they conform (just as one may play a simple game "correctly" and yet not be able to formulate its rules explicitly). All this belongs in the domain of the psychology of learning, motivation, symbolic behavior, and the like.

In any case, the problems of intentionality, and hence the relations of the logical to the psychological (or physiological, computerological, "robotological") are fundamentally different from the enigma of the relations of sentience to the physical processes. Some philosophers feel that the central issue of the mind-body problems is that of intentionality (sapience); others see it in the problem of sentience; and still others in the puzzles of selfhood. Although I have focused my attention primarily on the sentience problem, I regard the others as equally important. But I must confess that, as before, the sapience and selfhood issues have always vexed me less severely than those of sentience.

Returning, then, to the sentience problem, there are some aspects even of intentionality and of selfhood which may well require phenomenological description. By a sort of "lateral" view of the act-object relation—as we take it by introspection of perceiving, thinking, desiring, willing, etc.—we may say that we are "aware" of the intentionality or "aboutness" of such states of consciousness. But just as *inference* (in contradistinction to deducibility or entailment) is a psychological process, so the awareness of intentionality is a mental episode; and, again, if physicalism holds, some equivalent or "successor" account will eventually be given of these processes or episodes in physical$_2$ terms. Just what specifically such accounts will be remains to be settled by the future findings of neurophysiology. Similarly, what have been called the "conceptual" relations—such as those of the qualities to one another in the topological phenomenology of degrees of similarity or dissimilarity—must have their counterparts in certain features of brain processes by which we achieve the discriminatory judgments that are finally expressible in verbal or other types of responses. But we must not confuse that topological order with the causal order in which a given sensory episode occurs. Some sort of isomorphism is bound to prevail for each of these, i.e., between the phenomenal and certain features of the physical processes, but there will be striking differences between the two types of order.

Analogously, the much-discussed problems of the "nature of a person", of the "unity of consciousness", of the "identity of the self", and perhaps also of Kant's "synthetic unity of apperception" cannot be solved on a purely phenomenological basis. The phenomenal data in this domain, e.g., those having to do with the continuity of memory, or the ever (really only often) present possibility of connecting our current experience with earlier experiences (and the expectation of later ones) can be explained only by embedding them—as formulated in their respective successor concepts—within their neurophysiological setting. The psychiatric cases of dual or multiple personality may well be plausibly accounted for in terms of the alternating dominance of subsystems within the total set of brain processes.

There is one group of extremely difficult philosophical issues in which I have reached no more than highly tentative conclusions. A conceptual clarification is urgently needed regarding the differences (if any) between various types of phenomenological descriptions, and perhaps also between phenomenology generally and introspective psychology. E. Husserl and many of his disciples have focused their primary attention on the pure intuition of "essences" and their (allegedly "a priori") relations. It seems to me that the results of phenomenological intuition (*Wesensschau*) and those of the ordinary language analysis (Wittgenstein, Austin, and their disciples, despite their rather diverse claims and emphases) coincide to a remarkable degree. What the phenomenologists consider as "a priori necessary and synthetic" propositions finds its counterpart (if not equivalent) in the "conceptual necessities" uncovered by the linguistic analysts. If, for example, the "logical" incompatibility of "determinates" (e.g., red and green) under one "determinable" (color) is "intuited" phenomenologically as an "internal" relation of qualia, the same situation is described by the Wittgensteinians as part of the "grammar" of color words. Although I reserve some doubts on either account,* it must first of all be recognized that this sort of "synthetic a priori" is "puny" and insignificant compared with the "grandiose" a priori of the rationalists and of Kant (in regard to space, time, and causality). Whether the minor ("puny") a priori should be explicated (reconstructed or construed) in terms of syntactical formation rules or in terms of "A-postulates" (meaning rules à la Carnap or Maxwell) is irrelevant in the present context. I surmise that the phenomenologically intuited incompatibilities (or necessities) may well be basic *psychological* limitations (or constraints) on imaginability or even conceivability, and, if so, they may well be neurophysiologically explainable (ultimately!).

Introspective psychology seems to me to produce statements of a logically mixed character. If I introspect my current mental states—be they sensations, emotions, moods, intentions, desires, volitions—and report about them (as in a psychoanalytic interview), the egocentric particulars (at least "I" and "now") are almost always part of my utterances. But if I state a psychological regularity that is introspectively ascertained (e.g., great excitement always subsides after some time), then I disregard (abstract from) my own case and offer a generalization. This seems different indeed from the results of phenomenological intuition regarding the "internal" relations of "essences." To the extent that the customary introspective psychology formulates its knowledge claims in the frame of the "manifest image," i.e., the commonsense view, of the world, it uses a combination of subjective-phenomenal with intersubjective-scientific (spatio-temporal-causal) concepts.

Now, as is generally admitted, the manifest image—useful as it is in everyday life—is logically unstable, in that it contains implicit inconsistencies, and in that it is severely limited in its explanatory (and predictive) power. Behaviorism was and still is one remarkably successful way of securing consistency as well as explanatory and predictive power. I am here, of course, referring to *methodological* behaviorism (in contradistinction to logical or radical behaviorism which either denies or declares as meaningless purely mentalistic assertions). Relying on my foregoing remarks, I would say that as soon as the *peripher-*

* Consider, for example, *sweet* and *sour* as determinates under the determinable *taste quality*. These are clearly combinable and hence compatible, as is demonstrated by the taste of lemonade, or of sweet pickles!

alistic type of behaviorism (as, for example, in the outstanding work and basic orientation of B. F. Skinner) is supplemented by theories about the *central* states and processes within the organism, and especially in its nervous system, it is on its way to the kind of physicalism which forms the frame-hypothesis of the present philosophical analysis. What happens in this "great transformation" is the replacing of most (or all) concepts of the solipsistic (egocentric) perspective as well as of the manifest image (still suffused with subjectivistic features) by a completely intersubjective account. This has been seen, but expressed far too obscurely, even by the existentialists (e.g., Martin Buber), when they speak of the shift from the "I-Thou perspective" to the "It perspective" of impersonal, objective cognition. I have already discussed the "successor" concepts of the egocentric particulars; but along with them the radical objectification applies also to the experienced passage ("flow" or "flux") of time; the difference between past, present, and future; the "purposive" description of human action; the "intervention" (Collingwood) notion of cause; the value-impregnated notions of moral responsibility, freewill, and the "self." But again, nothing is "explained away"—all these features are merely subjected to a redescription in a thoroughly "detached" objective framework. The clamor about the "cleavage in our culture" between the sciences and the humanities may well be understood in terms of the shift in "perspective." There are not two different sorts of reality, but there are two ways of providing a conceptual frame for its description. In fact, at least so it seems to me, there are a great many "perspectives" or frames—the extremes being the purely egocentric as the "lower limit" and the completely $physical_2$ account as the "upper limit." In between are the many halfway (or part-way) houses of the possible manifest images. It is a good exercise for analytic philosophers to make explicit, in special reconstructions, the conceptual frame of each of these "perspectives."

While I should prefer not to irritate my tough-minded readers by waxing "metaphysical," I am tempted to say that the egocentric account, with its direct "labeling" of the qualities of experience, confronts *Being* (shudders?) as immediately as is possible in this world of ours, whereas all scientific accounts, owing to their quality-modality-invariance, deal with Being only indirectly and structurally.

The mistake criticized by the brilliant positivist ("empirio-criticist") R. Avenarius as an illegitimate "introjection"* of subjective experience into another person's body (head, brain) can easily be avoided if we adopt either the radically egocentric or the completely physicalistic account. The halfway houses of the manifest images inevitably lead to inconsistencies (or at least paradoxes, aporias) engendered by category mistakes. Of course, I agree that in the world of everyday life (is this really the *Lebenswelt* of the phenomenologists?) we understand each other quite well, even though the language of the "manifest image" is, strictly speaking, inconsistent. For the purposes of common communication the "introjection" is harmless. I suspect that even slightly sophisticated commonsense persons do not literally introject raw feels into the other fellow's brain. He would never expect to find images, sensations, emotions, and moods literally in the brain (not even in his own—if he were to examine it autocerebroscopically). In the manifest conception the brain is simply the "bloody mess" of neural tissue that one would come to see when opening a man's cranium. The region of (phenomenal) space in which the brain-as-seen appears is already occupied by the grayish-red stuff. (Leibniz was already quite clear about this.) Hence, even the "man-in-the-street," if he is not completely stupid, will understand the ascription of raw feels to other persons in the sense of the counterfactual (really counter-identical) proposition: "If I were the other fellow, and if I were in his particular momentary situation, I would have such and such experiences." Thus even commonsense is able to avoid (in this way) the paradoxes of introjection. This holds equally for one's own case in that in the autocerebroscopic situation one would have simultaneously some musical experiences, for example, along with some visual experiences which would furnish the confirming data for assertions about one's own cerebral processes.

It has been tempting to several thinkers to view the categorial incompatibility of the phenomenal and the physical language as a sort of "complementarity" analogous to the one of the Copenhagen interpretation of Quantum Mechanics. I think the analogy, though suggestive, is rather weak. Aside from the question of the tenability of the complementarity doctrine even in theoretical physics, I think that

*The reader should not confuse the meaning of this term with what Freud meant by "introjection"; Freud's concept refers to the incorporating of, for example, the father image, in our superego.

the mutual exclusiveness of the phenomenal and physical conceptual frames is to be explicated by the *logic* (semiotic) of the respective categories—and not as a formulation of a feature of the world. Correspondence rules connecting physical with phenomenal terms, however, are "crosscategorial" (see Cornman, 46, 48). They should be formulated in a semiotic (semantic-pragmatic) metalanguage.

My tentative conclusion is, as may be evident by now, that "introjection" rests on the category mistake of mixing purely egocentric language with objective (intersubjective) language. The basic difference, let me repeat, is that between direct labeling and indirect description (based on the nomological net provided by a theory). Nevertheless the ascription of raw feels to other persons is achieved in the scientific language by the ascription in terms of successor concepts of a specific "structure" in the conceptual network of $physical_2$ science to a certain region (of $physical_2$) space-time. Whatever seems to be missing is provided by the above-mentioned counterfactuals (or counter-identicals), and their own peculiar emotive significance.

The pictorial-emotional significance that provides a peculiar "root flavor" for the concepts of the physical sciences is, as I have tried to show, cognitively irrelevant—important though it may be heuristically and didactically. But the meaning of subjective-phenomenal concepts (in the egocentric perspective) definitely involves their "root flavor." It does not matter that this turns out to be completely trivial when formulated semantically (e.g., " 'red' designates red"; " 'warm' designates warm"; etc., etc.). These are the "language entry rules" in Wilfrid Sellars' formulation. It should be noted that I said that the meaning of purely phenomenal terms "involves" their "root flavor"; I did not say that it specifies their meaning completely even in the phenomenal language. Equally important is the logical locus of phenomenal terms in the structure of cognate terms, representing the place of a given quale among more or less similar qualia within the respective modality, and the place of one modality among the others.

One of the most important tasks yet to be done that will lead toward a more complete solution of the sentience problems is a precise logical analysis of the relation of the various phenomenal "spaces" (visual, tactual, kinesthetic, auditory, etc.) to physical space. Here the by now classical suggestions contained in the work of Schlick, Köhler, Russell, and Ruyer may well provide a useful starting point. Norbert Bischof

(20) seems to me to have contributed, quite recently, most fruitfully to this endeavor.

Another important task (only adumbrated above) concerns the exact syntactical, semantical, and pragmatic characterization of the (or a) phenomenal language. I have not been able to come up with anything better than the solipsistic reconstruction in terms of an *event* ontology in which phenomenal predicates (elementaristic or configurational, as the case may be) are ascribed to moments ("specious presents") of phenomenal time. Although I am by no means sure, I am inclined to think that this sort of (artificial) reconstruction might allow for a purely subjective language; all genuinely intersubjective or physical concepts would be kept out of it. But data as described in such a phenomenal language would provide the "ultimate" testing ground for all intersubjective propositions. (Of course this will require some sort of correspondence rules.)

At the risk of misleading the reader by a very weak and distant analogy, I suggest that the relation of the egocentric to the intersubjective account of the world may be compared with the relation of a geocentric to a heliocentric account of the kinematics of the planetary system. Just as the looplike or retrograde motions of the geocentric description disappear once the "Copernican turn" has been accomplished, so the directly given qualities and the "privileged" egocentric terms disappear in the intersubjective account of the physicalistic conception. And just as it is otiose (if not preposterous) to ask "Where are the epicycles in the heliocentric* description?" so the question "Where are the experienced subjective qualities in the scientific description of the world?" is equally inappropriate. These phenomenal qualities *are* described, but in a thoroughly different way, in the "transformed" account of intersubjective science. I think a basically similar approach will resolve the problems arising from the currently fashionable emphasis on the difference between "actions" and "movements." Actions as conceived intensionally (because of their "intentionality") belong to the manifest image of the world (suffused with egocentric

* I refer here not to the historical stages of the Ptolemaic and Copernican systems, but to a modern kinematic description of the geocentric, and heliocentric types, both brought up to date as regards the precise distances, orientation, etc. of the sun and the planets. (These are two systems, of course, logically equivalent, differing only in formal simplicity. But there is, by contrast, no L-equivalence between the egocentric and the physicalistic descriptions.)

significance). They are represented in a radically transformed manner —indeed as "movements"—or if this term is burdened with a pejorative connotation, as "processes" occurring in the interaction of organisms with one another and with their environment, all conceived ultimately in terms of $physical_2$ concepts. Fortunately the most promising endeavors in current theoretical psychology pay no attention to the "ordinary language philosophies of mind." While I grant that intuitive, empathetic, introspective, and phenomenological approaches are heuristically valuable, I doubt that they can contribute more than a technique of arriving at hypotheses which then still have to be tested (confirmed or disconfirmed) by the usual methods of intersubjective science. The ordinary language approach, though often phenomenologically perceptive, is fraught with the dangers of a regression to the sort of commonsense psychology which is contained in the "intuitive psychological understanding" that any person of some experience possesses anyway. This is the "psychology" used quite effectively in the practical affairs of diplomats, ministers, politicians, businessmen, parents, nursemaids, and fishwives. There are few surprises, and hardly anything that could be incorporated in, for example, the theory of motivation.

Looking back to the antimetaphysics of the logical positivists (or even to Popper's demarcation of scientific from metaphysical propositions) I now feel it does not matter much (except in "philosophical politics") whether such problems as those of the "reality of the external world," of "other minds," or of the "inverted spectrum" are regarded as metaphysical or as scientific. Carnap, I think with some plausibility, branded them (in the formulation in which he presented them) as pseudoproblems; but Popper regards them as meaningful but metaphysical questions. The sort of reasoning that conceives of, and argues for, mental states in other persons is, I have always admitted, an extreme and degenerate form of analogical reasoning. It depends on what one makes of those basic counterfactuals. I consider them as perfectly meaningful because their counterfactuality hinges upon certain fundamental natural limitations of direct testability. But once these limitations (the egocentric, the present moment predicament, etc.) are seen to be basic features of our universe—as indeed they appear in the best scientific and epistemological accounts—then perhaps the assorted aporias of philosophy and the "paradoxes" of existence will lose their traditional air of mystery, and a more enlightened philosophy will finally relieve us of those perennial perplexities.*

I realize only too painfully that the observations set down in this postscript are too sketchy and impressionistic to do more than, at best, provide suggestions for further, very much needed work in philosophical analysis. Short of writing a book (which I am not likely to do soon if ever) on *Mind and Its Place in Nature,*† I felt that it was better to present my current ideas on the occasion of the republication of my long essay of ten years ago than to remain silent. It is to my readers that I appeal for further concern with the mind-body perplexities, and for a charitable reception of my—possibly quite quixotic—ideas. In any case, it should be remembered that my entire discussion is predicated upon the scientific acceptability of (physical$_2$) physicalism. If future scientific research should lead to the adoption of one or another form of emergentism (or—*horribile dictu!*—dualistic interactionism), then most of my reflections will be reduced to the status of a logical (I hope not illogical!) exercise within the frame of an untenable presupposition. But since I now regard philosophical analysis as continuous with scientific research, I can only plead that we be permitted the procedure of trial and error, and of successive approximation in the predominantly philosophical endeavors just as we consider it entirely appropriate in the predominantly empirical but endless quest of scientific research.

* Keith Gunderson, in a forthcoming essay (78), has dealt with the vexatious predicaments both of "privileged access" and of "barred access" in a brilliant and highly original manner. I think he has succeeded in giving the most adequate intersubjective account of the puzzling asymmetries connected with subjectivity.

† The two remarkable books with this title, by C. D. Broad and Durant Drake, respectively, appeared in 1925! So, perhaps someone should try, on the current level of analytic sophistication, to bring the analysis scientifically and philosophically up to date.

Selected New References

Having given a much-appreciated, ample bibliography in the earlier essay, I present again a long list of items that I have found interesting, relevant, and controversial. However, the "book and article explosion" since 1957 is quantitatively so overwhelming that even this long list is unavoidably incomplete. I hope that I have not overlooked some genuinely important publications.

1. Abelson, R. "Persons, P-Predicates, and Robots," *American Philosophical Quarterly*, 3:306–311 (1966).
2. Agassi, J. "Sensationalism," *Mind*, 75:1–24 (1966).
3. Albritton, R. "On Wittgenstein's Use of the Term 'Criterion,'" *Journal of Philosophy*, 56:845–857 (1959).
4. Aldrich, V. C. "Behavior, Simulating and Nonsimulating," *Journal of Philosophy*, 63:453–456 (1966).
5. Anderson, A. R. (ed.). *Minds and Machines*. Englewood Cliffs, N.J.: Prentice-Hall, 1964.
6. Arbib, M. A. *Brains, Machines, and Mathematics*. New York: McGraw-Hill Book Co., 1964.
7. Armstrong, D. M. *A Materialistic Theory of Mind*. (Forthcoming)
8. Armstrong, D. M. *Bodily Sensations*. London: Routledge & Kegan Paul, 1962.
9. Armstrong, D. M. "Is Introspective Knowledge Incorrigible?" *Philosophical Review*, 72:417–432 (1963).
10. Armstrong, D. M. "The Nature of Mind," *Arts: Proceedings of the Sydney University Arts Association*, 3:37–48 (1966).
11. Armstrong, D. M. *Perception and the Physical World*. New York: The Humanities Press, 1961.
12. Aune, B. "Feelings, Moods, and Introspection," *Mind*, 72:187–208 (1963).
13. Aune, B. *Knowledge, Mind and Nature*. New York: Random House, 1967.
14. Aune, B. "Feigl on the Mind-Body Problem," in P. K. Feyerabend and G. Maxwell (eds.), *Mind, Matter, and Method*. Minneapolis: Univ. of Minnesota Press, 1966.
15. Aune, B. "On the Complexity of Avowals," in Max Black (ed.), *Philosophy in America*. Ithaca, N.Y.: Cornell Univ. Press, 1965.
16. Ayer, A. J. *The Concept of a Person and Other Essays*. New York: St. Martin's Press, 1963; London: Macmillan & Co., 1963.
17. Ayer, A. J. (ed.). *Logical Positivism*. New York: The Free Press, 1959.
18. Ayer, A. J. "Verification and Experience," in A. J. Ayer (ed.), *Logical Positivism*. New York: The Free Press, 1959.
19. Beloff, J. *The Existence of Mind*. New York: The Citadel Press, 1962.
20. Bischof, N. "Erkenntnistheoretische Grundlagenprobleme der Wahrnehmungspsychologie," *Handbuch der Psychologie*. Göttingen, Germany: Verlag für Psychologie, Dr. C. J. Hogrefe, 1966.
21. Bischof, N. "Psychophysik der Raumwahrnehmung," *Handbuch der Psy-*

chologie. Göttingen, Germany: Verlag für Psychologie, Dr. C. J. Hogrefe, 1966.
22. Blanshard, B. *Reason and Analysis*. LaSalle, Ill.: Open Court Pub. Co., 1962.
23. Brodbeck, M. "Meaning and Action," *Philosophy of Science*, 30:309–324 (1963).
24. Brodbeck, M. "Mental and Physical: Identity versus Sameness," in P. K. Feyerabend and G. Maxwell (eds.), *Mind, Matter, and Method*. Minneapolis: Univ. of Minnesota Press, 1966.
25. Brody, N., and Paul Oppenheim. "Tensions in Psychology between the Methods of Behaviorism and Phenomenology," *Psychological Review*, 73: 295–305 (1966).
26. Brown, R. *Explanation in Social Science*. Chicago: Aldine Pub. Co., 1963.
27. Buck, R. "Non-Other Minds," in R. J. Butler (ed.), *Analytical Philosophy*. New York: Barnes & Noble, 1962.
28. Bunge, M. *The Critical Approach to Science and Philosophy*. Glencoe, Ill.: The Free Press, 1964.
29. Cantril, H. "Sentio, Ergo Sum: 'Motivation' Reconsidered," *Journal of Psychology*, 65:91–107 (1967).
30. Cantril, H., and W. K. Livingston. "The Concept of Transaction in Psychology and Neurology," *Journal of Individual Psychology*, 19:3–16 (1963).
31. Capitan, W. H., and D. D. Merrill (eds.). *Metaphysics and Explanation*. Pittsburgh: Univ. of Pittsburgh Press, 1965.
32. Carnap, R. *Philosophical Foundations of Physics*. New York: Basic Books, 1966.
33. Carnap, R. "Psychology in Physical Language," in A. J. Ayer (ed.), *Logical Positivism*. New York: The Free Press, 1959.
34. Caspari, E. "On the Conceptual Basis of the Biological Sciences," in R. G. Colodny (ed.), *Frontiers of Science and Philosophy*. Pittsburgh: Univ. of Pittsburgh Press, 1962.
35. Castaneda, H. N. "Criteria, Analogy, and Knowledge of Other Minds," *Journal of Philosophy*, 59:533–546 (1962).
36. Castaneda, H. N. (ed.). *Intentionality, Minds, and Perception*. Detroit: Wayne State Univ. Press, 1967.
37. Castaneda, H. N. "The Private Language Argument," in C. D. Rollins (ed.), *Knowledge and Experience*. Pittsburgh: Univ. of Pittsburgh Press, 1963.
38. Castell, A. *The Self in Philosophy*. New York: The Macmillan Co., 1965.
39. Chappell, V. C. (ed.). *The Philosophy of Mind*. Englewood Cliffs, N. J.: Prentice-Hall, 1962.
40. Chisholm, R. M. *Perceiving: A Philosophical Study*. Ithaca, N.Y.: Cornell Univ. Press, 1957.
41. Chisholm, R. M. (ed.). *Realism and the Background of Phenomenology*. Glencoe, Ill.: The Free Press, 1960.
42. Chisholm, R. M. *Theory of Knowledge*. Englewood Cliffs, N.J.: Prentice-Hall, 1966.
43. Coburn, R. C. "Recent Work in Metaphysics," *American Philosophical Quarterly*, 1:204–220 (1964).
44. Coburn, R. C. "Shaffer on the Identity of Mental States and Brain Processes," *Journal of Philosophy*, 60:89–92 (1963).
45. Cook, J. W. "Wittgenstein on Privacy," *Philosophical Review*, 74:281–314 (1965).
46. Cornman, J. W. "The Identity of Mind and Body," *Journal of Philosophy*, 59:486–492 (1962).
47. Cornman, J. W. "Linguistic Frameworks and Metaphysical Questions," *Inquiry*, 7:129–142 (1964).

48. Cornman, J. W. *Metaphysics, Reference and Language*. New Haven and London: Yale Univ. Press, 1966.
49. Craik, K. J. W. *The Nature of Psychology*. New York: Cambridge Univ. Press, 1966.
50. Culbertson, J. T. *The Mind of Robots*. Urbana, Ill.: Univ. of Illinois Press, 1963.
51. Drange, T. *Type Crossings*. The Hague and Paris: Mounton & Co., 1966.
52. Ducasse, C. J. *The Belief in a Life after Death*. Springfield, Ill.: Thomas Pub. Co., 1961.
53. Eccles, J. C. (ed.). *Brain and Conscious Experience*. New York: Springer-Verlag, 1966.
54. Eccles, J. C. *The Brain and the Unity of Conscious Experience*. London: Cambridge Univ. Press, 1965.
55. Ekstein, R. "The Psychoanalyst and His Relationship to the Philosophy of Science," in P. K. Feyerabend and G. Maxwell (eds.), *Mind, Matter, and Method*. Minneapolis: Univ. of Minnesota Press. 1966.
56. Elsasser, W. M. *Atom and Organism*. Princeton, N.J.: Princeton Univ. Press, 1966.
57. Feigl, H. "Critique of Intuition According to Scientific Empiricism," *Philosophy East and West*, 8:1–16 (1958).
58. Feigl, H. "From Logical Positivism to Hypercritical Realism," *Proceedings of the XIIth International Congress of Philosophy* (Mexico City), 5:427–436 (1964).
59. Feigl, H. "Matter Still Largely Material," *Philosophy of Science*, 29:39–46 (1962).
60. Feigl, H. "Mind-Body, Not a Pseudoproblem," in S. Hook (ed.), *Dimensions of Mind*. New York: New York Univ. Press, 1960. Reprinted in Jordan Scher (ed.), *Theories of the Mind*. New York: The Free Press, 1960.
61. Feigl, H. "Other Minds and the Egocentric Predicament," *Journal of Philosophy*, 55:978–987 (1958).
62. Feigl, H. "Philosophical Embarrassments of Psychology," *American Psychologist*, 14:115–128 (1959). Reprinted in *Psychologische Beiträge*, 6:340–364 (1962).
63. Feigl, H. "Philosophy of Science," in Richard Schlatter (ed.), *Humanistic Scholarship in America*. Englewood Cliffs, N.J.: Prentice-Hall, 1964.
64. Feigl, H. "Physicalism, Unity of Science and the Foundations of Psychology," in P. A. Schilpp (ed.), *The Philosophy of Rudolf Carnap*. LaSalle, Ill.: Open Court Pub. Co., 1963.
65. Feigl, H. "The Power of Positivistic Thinking: An Essay on the Quandaries of Transcendence," *Proceedings and Addresses of the American Philosophical Association*, 36:21–41 (1963).
66. Feigl, H. "Why Ordinary Language Needs Reforming" (with G. Maxwell), *Journal of Philosophy*, 58:488–498 (1961).
67. Feyerabend, P. K. "How to Be a Good Empiricist—a Plea for Tolerance in Matters Epistemological," in Bernard Baumrin (ed.), *Philosophy of Science: The Delaware Seminar*, Vol. 2. New York: John Wiley & Sons, 1963.
68. Feyerabend, P. K. "Materialism and the Mind-Body Problem," *Review of Metaphysics*, 17:49–66 (1963).
69. Feyerabend, P. K. "Mental Events and the Brain," *Journal of Philosophy*, 60:295–296 (1963).
70. Feyerabend, P. K. "Problems of Empiricism," in R. G. Colodny (ed.), *Beyond the Edge of Certainty*. Englewood Cliffs, N.J.: Prentice-Hall, 1965.
71. Fink, D. G. *Computors and the Human Mind*. Garden City, N.Y.: Doubleday & Co., 1966.

72. Flew, A. (ed.). *Body, Mind, and Death.* New York: The Macmillan Co., 1964.
73. Fodor, J. A. "Explanations in Psychology," in Max Black (ed.), *Philosophy in America.* Ithaca, N.Y.: Cornell Univ. Press, 1965.
74. Gale, R. M. "The Egocentric Particular and Token-Reflexive Analyses of Tense," *Philosophical Review,* 73:213–228 (1964).
75. Garnett, A. C. "Body and Mind—the Identity Thesis," *Australasian Journal of Philosophy,* 43:77–81 (1965).
76. George, F. H. *The Brain as a Computer.* New York: Pergamon Press, 1962.
77. Grossmann, R. *The Structure of Mind.* Madison: Univ. of Wisconsin Press, 1965.
78. Gunderson, K. "Asymmetries and Mind-Body Perplexities," in H. Feigl and G. Maxwell (eds.), *Minnesota Studies in the Philosophy of Science,* Vol. IV. (Forthcoming)
79. Gunderson, K. "Cybernetics," in Paul Edwards (ed.), *Encyclopedia of Philosophy.* New York, 1967.
80. Gunderson, K. "Cybernetics and Mind-Body Problems." Presented at the meetings of the American Philosophical Association, Eastern Division, New York, December 29, 1965. (Forthcoming in *Methodos*)
81. Gunderson, K. "Descartes, La Mettrie, Language, and Machines," *Philosophy,* 39:193–222 (1964).
82. Gunderson, K. "The Imitation Game," in Alan Ross Anderson (ed.), *Minds and Machines.* Englewood Cliffs, N. J.: Prentice-Hall, 1964. Reprinted from *Mind.*
83. Gustafson, D. F. (ed.). *Essays in Philosophical Psychology.* Garden City, N.Y.: Doubleday & Co., 1964.
84. Hampshire, S. (ed.). *Philosophy of Mind.* New York: Harper & Row, 1966.
85. Hampshire, S. *Thought and Action.* London: Basil Blackwell, 1958.
86. Handy, R. *Methodology of the Behavioral Sciences.* Springfield, Ill.: Thomas Pub. Co., 1964.
87. Harris, E. E. *The Foundations of Metaphysics in Science.* New York: The Humanities Press, 1965.
88. Hawkins, D. *The Language of Nature.* San Francisco: W. H. Freeman & Co., 1964.
89. Hayek, F. A. *The Sensory Order.* Chicago: Univ. of Chicago Press, 1963.
90. Hesse, M. B. *Models and Analogies in Science.* London: Sheed & Ward, 1963.
91. Hillman, D. J. "On Grammars and Category-Mistakes," *Mind,* 22:224–234 (1963).
92. Hirst, R. J. (ed.). *Perception and the External World.* New York: The Macmillan Co., 1965.
93. Hook, S. (ed.). *Dimensions of Mind.* New York: New York Univ. Press, 1960.
94. Jonas, H. *The Phenomenon of Life.* New York: Harper & Row, 1966.
95. Kenny, A. *Action, Emotion and Will.* New York: The Humanities Press, 1963.
96. Kim, J. "On the Psycho-Physical Identity Theory," *American Philosophical Quarterly,* 3:227–235 (1966).
97. Kneale, W. C. "Broad on Mental Events and Epiphenomenalism," in P. A. Schilpp (ed.), *The Philosophy of C. D. Broad.* New York: Tudor Pub. Co., 1959.
98. Koch, S. "Psychology and Emerging Conception of Knowledge as Unitary," in T. W. Wann (ed.), *Behaviorism and Phenomenology.* Chicago: Univ. of Chicago Press, 1964.
99. Köhler, W. "A Task for Philosophers," in P. K. Feyerabend and G. Maxwell

(eds.), *Mind, Matter, and Method.* Minneapolis: Univ. of Minnesota Press, 1966.
100. Körner, S. *Experience and Theory.* New York: The Humanities Press, 1965; London: Routledge & Kegan Paul, 1965.
101. Kraft, V. *Erkenntnislehre.* Vienna: Springer-Verlag, 1960.
102. Lehrer, K. (ed.). *Freedom and Determinism.* New York: Random House, 1966.
103. Levi, I., and S. Morgenbesser. "Belief and Disposition," *American Philosophical Quarterly,* 1:221–232 (1964).
104. Lewis, D. K. "An Argument for the Identity Theory," *Journal of Philosophy,* 63:17–25 (1966).
105. Libet, B. "Cortical Activation in Conscious and Unconscious Experience," *Perspectives in Biology and Medicine,* 9:77–86 (1965).
106. Madden, E. H. "Problems in the Philosophy of Mind," *Southern Journal of Philosophy,* 4:33–40 (1966).
107. Madden, E. H. *Philosophical Problems of Psychology.* New York: Odyssey Press, 1962.
108. Malcolm, N. "Behaviorism as a Philosophy of Psychology," in T. W. Wann (ed.), *Behaviorism and Phenomenology.* Chicago: Univ. of Chicago Press, 1964.
109. Malcolm, N. "Explaining Behavior," *Philosophical Review,* 76:97–104 (1967).
110. Malcolm, N. "Knowledge of Other Minds," *Journal of Philosophy,* 55:969–978 (1958).
111. Mandelbaum, M. *Philosophy, Science and Sense Perception.* Baltimore: Johns Hopkins Press, 1964.
112. Margolis, J. "Brain Processes and Sensations," *Theoria, a Swedish Journal of Philosophy and Psychology,* 31:133–138 (1965).
113. Martin, C. B., and Max Deutscher. "Remembering," *Philosophical Review,* 75:161–196 (1966).
114. Matson, W. I. "Why Isn't the Mind-Body Problem Ancient?" in P. K. Feyerabend and G. Maxwell (eds.), *Mind, Matter, and Method.* Minneapolis: Univ. of Minnesota Press, 1966.
115. Meehl, P. E. "The Compleat Autocerebroscopist: A Thought-Experiment on Professor Feigl's Mind-Body Identity Thesis," in P. K. Feyerabend and G. Maxwell (eds.), *Mind, Matter, and Method.* Minneapolis: Univ. of Minnesota Press, 1966.
116. Miles, T. R. "The 'Mental'-'Physical' Dichotomy," *Aristotelian Society Proceedings,* 64:71–84 (1964).
117. Morgenbesser, S., and J. Walsh (eds.). *Free Will.* Englewood Cliffs, N.J.: Prentice-Hall, 1962.
118. Mundle, C. W. K. "Private Language and Wittgenstein's Kind of Behaviorism," *Philosophical Quarterly,* 16:35–46 (1966).
119. Nagel, E., and R. Brandt (eds.). *Meaning and Knowledge.* New York: Harcourt, Brace & World, 1965.
120. Nagel, E. *The Structure of Science.* New York: Harcourt, Brace & World, 1961.
121. Nagel, T. "Physicalism," *Philosophical Review,* 74:339–356 (1965).
122. Nelson, J. O. "An Examination of D. M. Armstrong's Theory of Perception," *American Philosophical Quarterly,* 1:154–160 (1964).
123. Passmore, J. *A Hundred Years of Philosophy,* second edition. New York: Basic Books, 1966.
124. Pepper, S. C. *Concept and Quality.* Berkeley: Univ. of California Press, 1967.

125. Perkins, M. "Emotion and the Concept of Behavior," *American Philosophical Quarterly*, 3:291–298 (1966).
126. Perkins, M. "Two Arguments against a Private Language," *Journal of Philosophy*, 57:443–459 (1965).
127. Pollock, J. L. "Criteria and Our Knowledge of the Material World," *Philosophical Review*, 76:28–60 (1967)
128. Popper, K. R. *Conjectures and Refutation*. New York: Basic Books, 1962.
129. Popper, K. R. *Of Clouds and Clocks*. Arthur Holly Compton Memorial Lecture presented at Washington University. St. Louis, Mo.: Washington University, 1965.
130. Popper, K. R. *The Logic of Scientific Discovery*. New York: Basic Books, 1959.
131. Pratt, C. C. "Free Will," in P. K. Feyerabend and G. Maxwell (eds.), *Mind, Matter, and Method*. Minneapolis: Univ. of Minnesota Press, 1966.
132. Price, H. H. "Appearing and Appearances," *American Philosophical Quarterly*, 1:3–19 (1964).
133. Price, H. H. "The Nature and Status of Sense-Data in Broad's Epistemology," in P. A. Schilpp (ed.), *The Philosophy of C. D. Broad*. New York: Tudor Pub. Co., 1959.
134. Putnam, H. "Brains and Behavior," in R. J. Butler (ed.), *Analytical Philosophy*, second series. New York: Barnes & Noble, 1965.
135. Putnam, H. "How Not to Talk about Meaning," in R. S. Cohen and M. W. Wartofsky (eds.), *Boston Studies in the Philosophy of Science*, Vol. II. New York: The Humanities Press, 1965.
136. Putnam, H. "Psychological Concepts, Explication and Ordinary Language," *Journal of Philosophy*, 54:95–100 (1957).
137. Putnam, H. "Robots: Machines or Artificially Created Life," *Journal of Philosophy*, 61:668–691 (1964).
138. Quinton, A. "The Soul," *Journal of Philosophy*, 59:393–409 (1962).
139. Rankin, K. W. "Referential Identifiers," *American Philosophical Quarterly*, 1:1–11 (1964).
140. Rescher, N. "The Revolt against Process," *Journal of Philosophy*, 59:410–417 (1962).
141. Rorty, R. "Mind-Body Identity, Privacy, and Categories," *Review of Metaphysics*, 19:24–54 (1965).
142. Routley, R., and V. Macrae. "On the Identity of Sensations and Physiological Occurrences," *American Philosophical Quarterly*, 3:87–110 (1966).
143. Russell, B. *My Philosophical Development*. New York: Simon & Schuster, 1959.
144. Scher, J. (ed.). *Theories of the Mind*. New York: The Free Press, 1962.
145. Schilpp, P. A. (ed.). *The Philosophy of Rudolf Carnap*. Library of Living Philosophers. LaSalle, Ill.: Open Court Pub. Co., 1963.
146. Schlick, M. "The Foundation of Knowledge," in A. J. Ayer (ed.), *Logical Positivism*. New York: The Free Press, 1959
147. Schrödinger, E. *Mind and Matter*. Cambridge: Cambridge Univ. Press, 1958.
148. Scriven, M. "The Limitations of the Identity Theory," in P. K. Feyerabend and G. Maxwell (eds.), *Mind, Matter, and Method*. Minneapolis: Univ. of Minnesota Press, 1966.
149. Scriven, M. "The Logic of Criteria," *Journal of Philosophy*, 56:857–868 (1959).
150. Scriven, M. *Primary Philosophy*. New York: McGraw-Hill Book Co., 1966.
151. Scriven, M. "Scientific Method: The Foundation of Psychology," *Psychology*. Boston: Allyn and Bacon, 1961.

152. Scriven, M. "Views of Human Nature," in T. W. Wann (ed.), *Behaviorism and Phenomenology*. Chicago: Univ. of Chicago Press, 1964.
153. Sellars, W. S. "The Identity Approach to the Mind-Body Problem," *Review of Metaphysics*, 18:430–451 (1965).
154. Sellars, W. S. "Notes on Intentionality," *Journal of Philosophy*, 61:655–665 (1964).
155. Sellars, W. S. "Philosophy and the Scientific Image of Man," in R. G. Colodny (ed.), *Frontiers of Science and Philosophy*. Pittsburgh: Univ. of Pittsburgh Press, 1962.
156. Sellars, W. S. *Science, Perception and Reality*. New York: The Humanities Press, 1963; London: Routledge & Kegan Paul, 1963.
157. Sellars, W. S. "The Refutation of Phenomenalism: Prolegomena to a Defense of Scientific Realism," in P. K. Feyerabend and G. Maxwell (eds.), *Mind, Matter, and Method*. Minneapolis: Univ. of Minnesota Press, 1966.
158. Sellars, W. S. "Theoretical Explanation," in Bernard Baumrin (ed.), *Philosophy of Science: The Delaware Seminar*, Vol. 2. New York: John Wiley & Sons, 1963.
159. Shaffer, J. A. "Persons and Their Bodies," *Philosophical Review*, 75:59–77 (1966).
160. Shaffer, J. A. "Could Mental States Be Brain Processes," *Journal of Philosophy*, 58:813–822 (1961).
161. Shaffer, J. A. "Recent Work on the Mind-Body Problem," *American Philosophical Quarterly*, 2:81–104 (1965).
162. Shapere, D. "Meaning and Scientific Change," in R. G. Colodny (ed.), *Mind and Cosmos*. Pittsburgh: Univ. of Pittsburgh Press, 1966.
163. Shwayder, D. S. *The Stratification of Behaviour*. New York: The Humanities Press, 1965.
164. Simon, H. A. "Thinking by Computers," in R. G. Colodny (ed.), *Mind and Cosmos*. Pittsburgh: Univ. of Pittsburgh Press, 1966.
165. Skinner, B. F. "Behaviorism at Fifty," in T. W. Wann (ed.), *Behaviorism and Phenomenology*. Chicago: Univ. of Chicago Press, 1964.
166. Smart, J. J. C. "Conflicting Views about Explanation," in R. S. Cohen and M. W. Wartofsky (eds.), *Boston Studies in the Philosophy of Science*, Vol. II. New York: The Humanities Press, 1965.
167. Smart, J. J. C. "Materialism," *Journal of Philosophy*, 60:651–662 (1963).
168. Smart, J. J. C. "Tensed Statements: A Comment," *Philosophical Quarterly*, 12:264 (1962).
169. Smart, J. J. C. *Philosophy and Scientific Realism*. New York: The Humanities Press, 1963; London: Routledge & Kegan Paul, 1963.
170. Smart, J. J. C. "The Identity Thesis—a Reply to Professor Garnett," *Australasian Journal of Philosophy*, 43:82–83 (1965).
171. Smythies, J. R. (ed.). *Brain and Mind*. New York: The Humanities Press, 1965; London: Routledge & Kegan Paul, 1965.
172. Spiegelberg, H. "Toward a Phenomenology of Experience," *American Philosophical Quarterly*, 1:325–332 (1964).
173. Stegmüller, W. *Hauptströmangen der Gegenwartsphilosophie*. Stuttgart: A. Kröner Verlag, 1965.
174. Stevens, S. S. "Quantifying the Sensory Experience," in P. K. Feyerabend and G. Maxwell (eds.), *Mind, Matter, and Method*. Minneapolis: Univ. of Minnesota Press, 1966.
175. Stoker, M. A. G. "Memories and the Private Language Argument," *Philosophical Quarterly*, 16:47–53 (1966).

176. Strang, C. "The Perception of Heat," *Aristotelian Society Proceedings*, 61: 239–252 (1961).
177. Strawson, P. F. *Individuals: An Essay in Descriptive Metaphysics*. Garden City, N.Y.: Doubleday & Co., 1963.
178. Swartz, R. J. *Perceiving, Sensing, and Knowing*. Garden City, N.Y.: Doubleday & Co., 1965.
179. Taylor, C. *The Explanation of Behavior*. New York: The Humanities Press, 1964.
180. Taylor, R. *Action and Purpose*. Englewood Cliffs, N.J.: Prentice-Hall, 1966.
181. Taylor, R. *Metaphysics*. Englewood Cliffs, N.J.: Prentice-Hall, 1963.
182. Thalberg, I. "Tenses and 'Now,'" *Philosophical Quarterly*, 13:3–15 (1963).
183. Thomas, G. "Abilities and Physiology," *Journal of Philosophy*, 61:321–327 (1964).
184. Thomson, J. J. "Private Languages," *American Philosophical Quarterly*, 1:20–31 (1964).
185. Thorpe, W. H. *Science, Man and Morals*. Ithaca, N.Y.: Cornell Univ. Press, 1966.
186. Turner, M. B. *Philosophy and the Science of Behavior*. New York: Appleton-Century-Crofts, 1967.
187. Van Peursen, C. A. *Body, Soul, Spirit: A Survey of the Body-Mind Problem*. London: Oxford Univ. Press, 1966.
188. Vesey, G. N. A. (ed.). *Body and Mind*. London: Allen & Unwin, 1964.
189. Vesey, G. N. A. *The Embodied Mind*. London: Allen & Unwin, 1965.
190. Wann, R. W. (ed.). *Behaviorism and Phenomenology*. Chicago: Univ. of Chicago Press, 1964.
191. Wellman, C. "Wittgenstein's Conception of a Criterion," *Philosophical Review*, 71:433–447 (1962).
192. White, A. R. *The Philosophy of Mind*. New York: Random House, 1967.
193. Williams, D. C. *Principles of Empirical Realism*. Springfield, Ill.: Thomas Pub. Co., 1966.
194. Wisdom, J. O. "Some Main Mind-Body Problems," *Aristotelian Society Proceedings*, 60:187–210 (1959–1960).
195. Wittgenstein, L. *The Blue and Brown Books*. New York: Harper & Bros., 1958.
196. Wittgenstein, L. *Philosophische Bemerkungen*. Oxford: Basil Blackwell, 1964.
197. Wolman, B. B. (ed.), and E. Nagel (consulting ed.). *Scientific Psychology*. New York: Basic Books, 1965.
198. Wright, S. "Biology and the Philosophy of Science," *Monist*, 48:265–290 (1964).
199. Wyburn, G. M. (ed.). *Human Senses and Perception*. Toronto: Univ. of Toronto Press, 1964.
200. Ziff, P. "The Simplicity of Other Minds," *Journal of Philosophy*, 62:575–584 (1965).

LATE ADDENDA

The following essays and articles (except the first two on the list) are forthcoming. They present a bold and new approach of realistic structuralism. I expect that this reconstruction—though in important points differing from my own—may yet offer the most illuminating solution of the sentience issue of the mind-body problems.

201. Maxwell, Grover. "The Ontological Status of Theoretical Entities," in H.

Feigl and G. Maxwell (eds.), *Minnesota Studies in the Philosophy of Science*, Vol. III. Minneapolis: University of Minnesota Press, 1962.

202. Maxwell, Grover. "Philosophy and the Causal Theory of Perception," *Graduate Review of Philosophy*, 6:9–21 (1964).
203. Maxwell, Grover. "Remarks on Perception and Theoretical Entities," in R. G. Colodny (ed.), *Pittsburgh Studies in the Philosophy of Science*, Vol. IV. Pittsburgh: Univ. of Pittsburgh Press, 1967.
204. Maxwell, Grover. "Scientific Methodology and the Causal Theory of Perception," in I. Lakatos and A. Musgrave (eds.), *Problems in the Philosophy of Science*. Amsterdam: North Holland Press, 1967.
205. Maxwell, Grover. "Reply" [to Professors Quine, Popper, Ayer, and Kneale], in I. Lakatos and A. Musgrave (eds.), *Problems in the Philosophy of Science*. Amsterdam: North Holland Press, 1967.
206. Maxwell, Grover. "Epistemology and Metaphysics as Natural Sciences," in P. A. Schilpp (ed.), *The Philosophy of Karl Popper*. LaSalle, Ill.: Open Court Pub. Co., forthcoming.

INDEX

Name Index

Subject Index

CPSIA information can be obtained
at www.ICGtesting.com
Printed in the USA
BVHW040106210220
572939BV00010B/66

9 780816 657599